A Research Agenda for Sustainable Tourism

Elgar Research Agendas outline the future of research in a given area. Leading scholars are given the space to explore their subject in provocative ways, and map out the potential directions of travel. They are relevant but also visionary.

Forward-looking and innovative, Elgar Research Agendas are an essential resource for PhD students, scholars and anybody who wants to be at the forefront of research.

Titles in the series include:

A Research Agenda for New Institutional Economics
Edited by Claude Ménard and Mary M. Shirley

A Research Agenda for Regeneration Economies
Reading City-Regions
Edited by John R. Bryson, Lauren Andres and Rachel Mulhall

A Research Agenda for Cultural Economics
Edited by Samuel Cameron

A Research Agenda for Environmental Management
Edited by Kathleen E. Halvorsen, Chelsea Schelly, Robert M. Handler, Erin C. Pischke and Jessie L. Knowlton

A Research Agenda for Creative Tourism
Edited by Nancy Duxbury and Greg Richards

A Research Agenda for Public Administration
Edited by Andrew Massey

A Research Agenda for Tourism Geographies
Edited by Dieter K. Müller

A Research Agenda for Economic Psychology
Edited by Katharina Gangl and Erich Kirchler

A Research Agenda for Entrepreneurship and Innovation
Edited by David B. Audretsch, Erik E. Lehmann and Albert N. Link

A Research Agenda for Financial Inclusion and Microfinance
Edited by Marek Hudon, Marc Labie and Ariane Szafarz

A Research Agenda for Global Crime
Edited by Tim Hall and Vincenzo Scalia

A Research Agenda for Transport Policy
Edited by John Stanley and David A. Hensher

A Research Agenda for Tourism and Development
Edited by Richard Sharpley and David Harrison

A Research Agenda for Housing
Edited by Markus Moos

A Research Agenda for Economic Anthropology
Edited by James G. Carrier

A Research Agenda for Sustainable Tourism
Edited by Stephen F. McCool and Keith Bosak

A Research Agenda for Sustainable Tourism

Edited by

STEPHEN F. McCOOL

Professor Emeritus, Department of Society and Conservation, University of Montana, USA

KEITH BOSAK

Professor of Nature-based Tourism and Recreation, Department of Society and Conservation, University of Montana, USA

Elgar Research Agendas

Cheltenham, UK • Northampton, MA, USA

Published by
Edward Elgar Publishing Limited
The Lypiatts
15 Lansdown Road
Cheltenham
Glos GL50 2JA
UK

Edward Elgar Publishing, Inc.
William Pratt House
9 Dewey Court
Northampton
Massachusetts 01060
USA

Paperback edition 2021

A catalogue record for this book
is available from the British Library

Library of Congress Control Number: 2019946817

This book is available electronically in the **Elgar**online
Social and Political Science subject collection
DOI 10.4337/9781788117104

ISBN 978 1 78811 709 8 (cased)
ISBN 978 1 78811 710 4 (eBook)
ISBN 978 1 80088 611 7 (paperback)

Typeset by Servis Filmsetting Ltd, Stockport, Cheshire
Printed and bound by CPI Group (UK) Ltd, Croydon, CR0 4YY

Contents

Figures

Tables

Contributors

Cherise Addinsall is a social scientist and specialist in sustainable tourism, and has been researching and supervising in sustainable livelihoods (and enhancing the livelihoods of women and marginalized people) in Australia and the South Pacific, with a particular focus on supporting sustainable development in tourism, agriculture and conservation initiatives.

Keith Bosak's research interests are broadly centered on the intersection of conservation and development, and as such he often studies nature-based tourism and sustainable tourism in the context of development and protected areas. He currently serves as a professor in the Department of Society and Conservation at the University of Montana. He also directs the International Seminar on Protected Area Management in conjunction with the US Forest Service International Programs.

Kelly S. Bricker is a professor and director of Parks, Recreation, and Tourism at the University of Utah, located in the College of Health. She specializes in nature's restorative benefits, amenity communities, tourism impacts, sustainable tourism and protected areas. With partners in OARS and her husband, she has developed an ecotourism operation, *Rivers Fiji.*

Lee K. Cerveny is a research social scientist and team leader of the People and Natural Resources Team in the US Forest Service Pacific Northwest Research Station. Cerveny works with federal land managers to develop public engagement approaches that rely on participatory mapping to identify special places and recreation patterns. Other research involves community-based collaborations, national forest partnerships, and understanding effects of homelessness on public lands.

Douglas Dalenberg is a Professor of Economics at the University of Montana specializing in public economics. His research focuses on using survey and non-survey data to better understand choices and behaviors of individuals.

Monika M. Derrien is a research social scientist with the USDA Forest Service's Pacific Northwest Research Station in Seattle, Washington. She is interested in the human dimensions of natural resource management, especially related to social, cultural and health aspects of outdoor recreation and tourism planning.

Stephen Espiner is a senior lecturer in parks, recreation and tourism at Lincoln University, New Zealand where his research focus is the human dimensions of national parks and protected area management. He has a particular interest in outdoor recreation, nature-based tourism and associated conservation, and community and visitor management issues, including sustainability and resilience in nature-based tourism destinations.

James Higham is a professor at the University of Otago (New Zealand), Visiting Professor of Sustainable Tourism at the University of Stavanger (Norway) (2008–20) and Jim Whyte Fellow, University of Queensland (Australia) (2016–19). With Professor Xavier Font (University of Surrey) he is co-editor of the *Journal of Sustainable Tourism*. His research interests generally address tourism, sustainable development and global environmental change.

Yu-Fai Leung is professor and director of graduate programs in the Department of Parks, Recreation and Tourism Management at North Carolina State University. His research supports sustainable tourism and recreation in protected areas, with special interests in recreation ecology, visitor use and impact monitoring, and applications of geospatial technology.

Johanna Loehr studies at the Griffith Institute for Tourism, and is a sustainability tourism industry professional. Her research interests are tourism and climate change, sustainable tourism and systems thinking. She is currently focusing on tourism and climate change adaptation in the South Pacific.

Stephen F. McCool is Professor Emeritus at the University of Montana. His research and instructional emphases focus on visitor and public use planning in protected areas, public engagement processes and protected area planning. He has published several books and authored over 200 publications. He often serves as a consultant to protected area agencies in different places in the world.

Teresa Cristina Magro-Lindenkamp is Professor of Protected Areas Management at the University of São Paulo (USP), Brazil. She is attached to the Forest Science Department of the Luiz de Queiroz College of Agriculture in Piracicaba, and has been teaching and working since 1987 in parks planning and environmental and social impacts of public use in protected areas.

Anna B. Miller is a recreation ecologist currently working with the US Forest Service Pacific Northwest Research Station. Her primary research focus is on human–wildlife interactions in the outdoor recreation context. She works to integrate social and ecological sciences in her research to contribute towards sustainable and resilient protected area management.

Zachary D. Miller is an assistant professor at Utah State University. His research focuses mainly on park and conservation area management, including visitor use management, human–wildlife relationships (interactions/conflict), and environmental communication (interpretation/education). More broadly, his research interests are related to social–ecological relationships, including conservation social sciences and other human dimensions of natural resources.

Gianna Moscardo has qualifications in applied psychology and sociology and joined the School of Business at James Cook University in 2002. Her qualifications in applied psychology and sociology support her research interests in understanding how communities and organizations perceive, plan for and manage tourism development opportunities, how tourists learn about and from their travel experiences, and how to design more sustainable tourism experiences. In 2014 she was appointed chair of the international BEST EN (Building Excellence in Sustainable Tourism Education Network) group.

Peter Newman is a professor and department chair in Recreation, Park, and Tourism Management at Penn State University. His research focuses on visitor use decision making in the context of protected areas management. His interests include visitor management in and regarding protected areas, soundscape/acoustic management in parks, transportation management and planning, efficacy of Leave No Trace practices, and health and well-being.

Caroline Orchiston is deputy director at the Centre for Sustainability at the University of Otago (New Zealand). Her research interests focus on natural hazards (including earthquakes, tsunami and climate-related hazards) and associated impacts on communities. Much of her research investigates mitigation and response to earthquake disasters from a community and tourism perspective, particularly post-earthquake recovery and organizational resilience.

Colter Pence is a natural resource specialist with the US Forest Service on the Flathead National Forest, Hungry Horse, Montana. She specializes in managing Wilderness, Wild and Scenic Rivers, trail systems, and other backcountry programs.

William L. Rice is a graduate student at Penn State University. His research focuses on personal and social outcomes associated with public lands recreation and the impact of protected areas' national and international significance. His projects are primarily set in large Western national parks.

Andrew Rylance is an independent consultant, specializing in environmental economics and conservation finance. He is currently the technical advisor to the Government of Seychelles-UNDP-GEF project on Protected Area Finance. He has over a decade of experience working in Africa and Europe. He is Research Fellow of the University of Johannesburg.

Jarkko Saarinen is a Professor of Human Geography at the University of Oulu, Finland, and Distinguished Visiting Professor (Sustainability Management) at the University of Johannesburg, South Africa. His research interests include tourism and development, sustainability and responsibility in tourism, tourism–community relations, tourism and climate change adaptation, and wilderness studies.

Daniel Scott is a professor and research chair at the University of Waterloo. He has worked extensively in the areas of climate change and the global tourism, including collaborations with the UN World Tourism Organization, UN Environment Programme, and the World Meteorological Organization. Dr. Scott has been a contributing author to the Intergovernmental Panel on Climate Change Third, Fourth, and Fifth Assessment Reports.

Anna Spenceley is a consultant focusing on sustainable tourism who has worked for over 20 years in developing countries on assignments promoting conservation and poverty reduction. She is chair of the IUCN's World Commission on Protected Areas (WCPA) Tourism and Protected Areas Specialist Group, and sits on the board of the Global Sustainable Tourism Council. She is also a senior research fellow with the University of Johannesburg, and an honorary fellow of the University of Brighton.

B. Derrick Taff is an assistant professor in Recreation, Park, and Tourism Management at Penn State University. His research strives to improve understanding of communication strategies aimed at influencing human and environmental health regarding protected areas. Specifically, he focuses on Leave No Trace-based behaviors, health and nature-based recreation, and human well-being as it pertains to natural sounds.

Jennifer Thomsen is an assistant professor in the Department of Society and Conservation at the University of Montana. Her research focuses on bridging personal backgrounds in natural science in wildlife and fisheries biology with social science in parks and conservation area management, in four main areas: (1) stakeholder collaboration associated with large landscape conservation, (2) sustainable tourism and protected area management, (3) the relationship between human and ecosystem health, and (4) the relationship between environmental learning and pro-environmental behavior.

Betty Weiler is a research professor at Southern Cross University. Her work focuses on sustainable tourism, with a particular interest in visitor management and the tourist experience. She has collaborated with numerous park management agencies and wildlife tourism attractions, and is widely known for her research and publications on tour guiding and heritage/nature interpretation.

Iree Wheeler is a master's student in the University of Montana College of Forestry and Conservation. Her research focuses on the development of indigenous conservation areas by tribes and first nations in Canada and the United States. She has also conducted research on visitor-use on the Flathead River System for the past two years.

Kathleen L. Wolf is a research social scientist with the College of the Environment at the University of Washington (Seattle). Her research focus is the human dimensions of urban forestry and urban ecosystems, particularly human health. Her research can be viewed at http://www.naturewithin.info; and the Green Cities: Good Health project at https://depts.washington.edu/hhwb/.

Preface

One of the great challenges the world faces today is dealing with the ever-growing complexity of environmental problems that are ever more difficult to resolve, not only in a technical sense but in a social and cultural sense as well. While some are solvable technically, they all have to be acceptable socially, for if a "solution" is not acceptable, it will not be implemented. One of those problem areas is tourism, which means all forms of tourism, not just those that are often characterized as sustainable.

The complexity of these problems demands that we better understand what their "roots" may be, what causes this complexity, what resolutions may be available in the context of the times, and how we may implement those resolutions. We point specifically to the notion of resolution, as with complex environments problems never stay solved because the context is dynamic. When a problem is resolved that means an agreement has been made about the way forward, not that an answer has been computed in a technical sense.

And yet we have to be careful, in the words of Russell Ackoff, not to be solving the wrong things because that just leads to us doing things wronger and wronger. Building understanding of the tourism system, or more correctly the tourism systems, for there are multiple ones, can help us improve our ability to do the right things, and if we are working on the right things, even if the outcomes are wrong, then we will get better and better over time. For example, focusing on efficiency is not always the right thing. Emphasizing equity may actually get us to our goal faster than pursuing efficiency.

Research helps us get better and better.

So this book is about the research needed to help the tourism field focus on the challenges we face in its numerous subcomponents. The need for policy-relevant research is great; research that helps policy makers and decision makers to carefully and sensitively advance an industry that we see has a high potential to enhance the quality of life for many on this planet. We have emphasized here many of the challenges confronting the idea of sustainable tourism, which we loosely define as a response to the question "What should tourism sustain?", which suggests an outcome rather than a kind of tourism. Up until the late 1990s, sustainable tourism was often viewed as the intersection of economic feasibility, social acceptability and ecological viability. The world has dramatically changed since that time, as we argue in our book *Reframing Sustainable Tourism* (2016, Springer). We now see sustainable tourism as a component of a larger social-ecological system. That view has resulted in more useful insights and questions.

The world is volatile, uncertain, complex and ambiguous (VUCA – a term coined in the 1990s by the U.S. military about the wars in the Balkans), itself a response to the invalid assumptions underlying decision making that it is predictable, linear, understandable and stable, as Kohl and McCool noted in their book *The Future has Other Plans* (2016, Fulcrum). This complex world is best addressed through systems thinking, which we do not define here but is discussed in Chapters 1–3. To the extent possible, we have asked authors to use a systems-thinking approach to frame their research agendas and challenges within the context of systems thinking. While the chapters are not placed in any specific order, we recommend a good reading of Chapters 1–3 prior to the others as those chapters are specifically concerned with various aspects of systems thinking.

These research challenges are many, and we have collected only a few stimulating essays. We believe the reader will be informed by their content. The chapters range from the direct and practical to the indirect and conceptual.

Each chapter looks through a window onto the larger scene of research of the relevant arena that both faculty and graduate students will find helpful. To some extent destination area managers and marketers may find these chapters useful in providing background when meeting with scientists about needs and approaches in developing useful information. We do not include chapters on sustainable and ecological economics, which are large subjects that are helpful, we believe, to sustainable tourism, because they are covered elsewhere, nor such subjects as sustainable materials and transportation.

Producing a book of this kind requires a large degree of collaboration and joint production of knowledge. We first thank the 26 authors who spent a large amount of their time voluntarily to organize and communicate their thoughts about the assignments. We also thank the people at Edward Elgar Publishing for suggesting this topic and editing the final product. Of course, we would not be able to produce any of this content if not for the hundreds, possibly thousands, of scientists who conducted the research upon which the book is based, and the many practitioners who not only needed this information and used it but thereby also provided the initial demand for the studies cited here. They were highly responsible for providing the stimulus to conduct the research summarized in this book. Ultimately, the desire for tourists to seek opportunities for transformative experiences led to the knowledge synthesized here.

We graciously recognize and thank our families for supporting this effort; if not for them, this volume would not have been produced.

Steve McCool and Keith Bosak
March 2019

1 Information needs for building a foundation for enhancing sustainable tourism as a development goal: an introduction

Stephen F. McCool

Introduction

Societies flourish when their members connect with and bond to their heritage, both cultural and natural. Understanding and appreciating that heritage is part of the human condition, so much so that otherwise we are lost as a species. This heritage and the connections it provides may be recreational (as when we seek outcomes such as adventure, challenge, escape, solitude and stress release), educational (such as learning about natural processes), cultural (such as appreciating how our societies have developed and understanding notable events of the past), spiritual, or utilitarian (such as harvesting resources for sustenance and shelter). The heritage values that support these connections, whether a remote landscape of the arctic, an historically significant cultural event in Asia, or a small protected forest in Eastern Europe, require careful stewardship if they are to be accessible to our grandchildren and theirs.

These connections are critical not only to our long-term survival as a species, but also in our day to day lives. But because of growing development and human population dynamics, the resources upon which we depend as a species are threatened. The International Union for the Conservation of Nature (IUCN) states "research is very clear about the human need for nature, but this situation is likely to worsen as the global population shifts from 54 per cent of people living in cities today to a forecast 70 per cent by 2050" (IUCN 2014).

Urbanization has many advantages; but one of the disadvantages is the risk of breaking connections between us and our inherited heritage. For example, concrete canyons of buildings, paved streets, and business-focused activity may lead to the loss of understanding of our environmental dependency for clean water, pure air, sustenance and shelter. And if connections between ourselves and our past are shattered, how will we learn to deal with the future? Around the globe there is substantial evidence of these connections being ruptured in contemporary society, in terms of declining biodiversity, in terms of the increasing vulnerability of poor people and in threats to water security. Building knowledge about these connections and implementing that knowledge thus becomes an essential endeavor of human development.

Providing careful stewardship of these connections is a goal, but a challenging one in the era of turbulence that characterizes the twenty-first century. Stewardship requires someone, or a group of people, to administer an area, monument or facility, to take care of that heritage, to ensure that the values it curates are indeed accessible to those living in the future as well as the current generation. Marketing destination areas also requires stewardship, although of a different character, one that involves creativity in connecting people with places, in developing interpretive programs that inspire people to learn about their heritage, in creating opportunities for people to have fun and be entertained, and in building opportunities for visitors to have transformative experiences. Solving problems, working with citizens and activists, developing partnerships and engaging in research are direct forms of stewardship as much as patrolling areas, interpreting natural and cultural sites, and maintaining them.

These special places (over 238,000 protected natural areas and many more hundreds of thousands of culturally significant areas as of 2019) are in high demand by the globe's population as places to visit, enjoy for recreation, appreciate and understand. As a result, tens of thousands of destination marketing organizations, development agencies, protected area management and cultural ministries, academics, managers and non-governmental organizations at scales from the local to the global are engaged in ensuring and managing access to them. That job requires an information and knowledge base to better understand what people seek in tourism, what the positive and negative social and biophysical consequences may be, and how tourism may be better managed. Providing opportunities for high-quality experiences is a particularly challenging task given the scale and the often vulnerable character of natural and cultural heritage to visitation that is unmanaged.

When you think about it, this is a pretty hefty responsibility for an industry, that in 2016 accounts, globally, for about 3.1 percent of the direct total global gross domestic product, employs 109 million people directly and about 193 million additional people indirectly. Moving forward, these figures are expected to rise significantly in the next ten years, with more than 381 million people whose jobs depend upon tourism (World Travel and Tourism Council (WTTC) 2017). The WTTC estimates that about three-quarters of all travel spending is for leisure.

The UN World Tourism Organization (UNWTO) expects travel to increase significantly through the year 2030. While 1.235 billion international arrivals were recorded in 2016, the UNWTO predicts 2030 arrivals at 1.8 billion, nearly 50 percent higher than the current level (UNWTO 2011, 2017). These figures reinforce the notion that making sustainability a high priority is the prudent thing to do. The UNWTO defines sustainable tourism as "Tourism that takes full account of its current and future economic, social and environmental impacts, addressing the needs of visitors, the industry, the environment and host communities" (UNWTO 2005). One of the challenges apparent in defining sustainable tourism in this manner is

Box 1.1

The small town of Kimmswick, Missouri in the United States provides an example of the challenges and opportunities faced in connecting people with its heritage. Kimmswick is located on the banks of the Mississippi River, just south of the major city of St. Louis, which in the mid-nineteenth century served as the gateway to westward expansion of Euro-American settlement of the U.S.A. Kimmswick itself has a record of human occupation going back approximately 12,000 years, but it was the westward expansion that led to its initial development as the community we recognize today. The town's population, fueled by extraction of iron ore exploded in the late nineteenth century to nearly 1,500 people, and then dropped dramatically to about the 150 or so that populate the community today.

Many of the nineteenth-century buildings have been preserved and contain small businesses catering to tourists, bed and breakfast operations and restaurants. The community is solely dependent on tourism for its economic viability. During the summer months, thousands of tourists, primarily from the local area will crowd the streets, businesses and restaurants enjoying the historical heritage primarily preserved in architecture.

And yet, all is not well. The Mississippi River occasionally floods the grounds of the Busch family (noted for their Budweiser brand of beer) mansion which itself is a destination and an equestrian center. Community leadership on the kind of tourism to attract occasionally falters. The community exists but is subject to changes in the economic situation originating hundreds of miles away, policies developed in the state and national capitals, and forces of climate change occurring at the global level, with little say in any of them. So, while we can say the community still exists, to a large extent as it was 150 years ago, has it been sustained? Is it resilient in the face of global-level trends and forces? Does it hold the vision, leadership and technical skills to survive a century of turbulence?

Kimmswick is broadly representative of thousands of small communities around the globe. They are struggling because of urbanization and abandonment of rural areas, grasping at tourism as the only economic opportunity they see to survive. And so, we ask the question, how can research help these communities develop a viable economic base so they become the vibrant community they want to be? This knowledge covers a wide variety of arenas, not limited to marketing information as generally conceived. But what kind of information would be most useful? Does the community have the capacity to use the information? What kind of capacities does the community need?

that it is often assumed to be a category of tourism, rather than looking at all tourism activity to confront sustainability.

Much of the focus on sustainability is development, making development more sustainable. In fact, 2017 was designated as the UN Year of Sustainable Tourism for Development. The primary goal of this recognition was to advance the UN's 2030 Agenda for Sustainable Development (UN 2015) through the use of tourism. That agenda included 17 goals for development ranging from ending poverty to ensuring access to energy (see Chapter 8 by Anna Spenceley and Andrew Rylance). Tourism will be a significant component of many integrated development strategies that can help make progress toward these goals.

Tourism is often viewed now as a method of development. Tourism activity is widespread around the globe; it has, as noted above, significant economic consequences; it can be an important source of foreign exchange, one that can be enhanced relatively easily; and requires less capital investment to create jobs than most other industries. Tourism can, as in other industries, if carefully managed, contribute to local labor income and taxes, encourage the development of local services, particularly health care, provide jobs that allow for upward mobility and otherwise enhance social capital (Shakya 2016). All these can lead to a more fulfilling quality of life for residents in local areas where tourism occurs. And at the same time, tourism carries with it the potential, whether involving visitation to natural or cultural sites, to generate revenues and other benefits that can be used for conservation of these values. Drumm (2007) notes that "Tourism certainly has an enormous potential to be a significant source of conservation finance for financially challenged protected area systems."

And yet, tourism also carries with it the potential for many negative social and environmental consequences, such as along scenically attractive coastlines and within fragile World Heritage Sites like Angkor Archaeological Park in Cambodia, as well as contributing to negative impacts on local cultures that are particularly thorny and difficult to manage. The presence of tourists, particularly large numbers relative to the local population, can disrupt normal societal functions, such as access to areas that provide important ecosystem services, involve culturally or spiritually significant values, or simply make access to locally popular destinations such as beaches or cafes difficult, and make highways and local roads congested (witness the rise of demonstrations against tourism in various European cities that underlie the notion of "overtourism").

Natural and cultural protected areas are not immune to the intense demand for the values they protect. La Tijuca National Park, which protects the mountains in and around Rio de Janeiro in Brazil, but also manages the "Christ the Redeemer" statue (the reason why most of the three million plus visitors go to the park), has had to institute a high-tech visitor use management system. Plitvice Lakes National Park in Croatia is considering something similar. Many other parks have also come to realize that the rapidly growing demand for the values they protect require active management of visitor use as well.

And all of this occurs within a context of turbulence: the twenty-first century is characterized by forces and processes occurring at both global and local levels, sometimes acting synergistically, but very often acting unpredictably and at all times producing volatility, complexity, change and uncertainty (Figure 1.1). Tourism activity and development occurs within this context: change is non-linear—many probable causes may lead to the same or similar effects; temporal delays of varying amounts typify the relationship between an action and its consequence; unanticipated effects ("surprises") often happen; consensus on action often has to be built; and integration of many disciplines is required to solve problems and build a sustainable future.

This book is about building the information and generating the knowledge needed to maintain connections between people and their heritage, both natural and

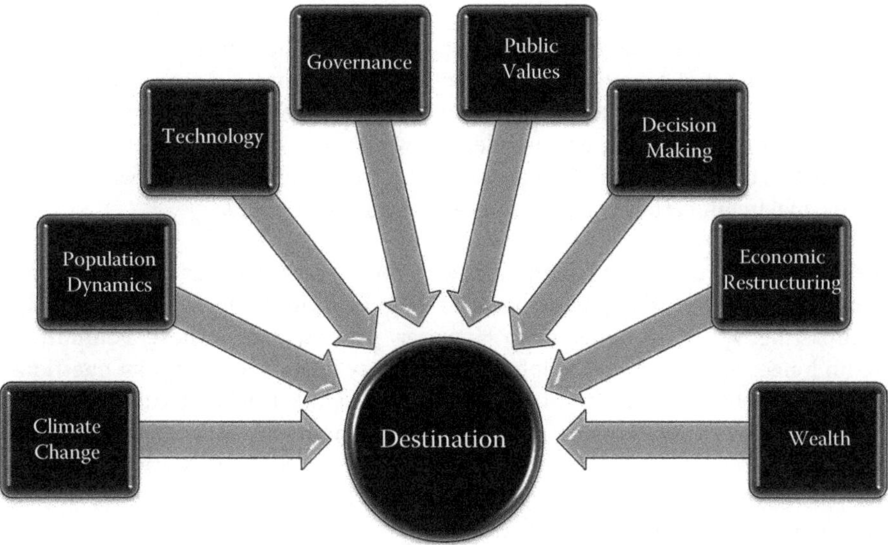

Figure 1.1 Large-scale forces acting upon tourism destinations

cultural. It focuses on connections that are primarily the foundation of tourism, experiences that people have when engaging this heritage, principally in a non-utilitarian way. The chapters identify information and research needs that decision makers can use to inform policy, planning, management and marketing. The book will be helpful primarily to academics and graduate students and other researchers in deciding on thesis and research projects, as well as to managers and granting bodies setting research priorities and making budget decisions. We note that to be useful, research must be relevant to tourism policy and management; we thus have emphasized points in an agenda that are more applied than theoretical, the "low hanging fruit" if you will, research that can be useful in a more immediate sense. Basic research is also a component, but of a more long-term nature.

By itself, research will not solve problems. It can only be applied by policy makers that hold the competency and the will to apply the research and who have the confidence of their constituencies. The research challenges mentioned here would be helpful to such policy makers and managers who recognize their limits and are willing to unite in manager–scientist partnerships that are particularly useful in turbulent contexts.

An era of turbulence

If there is anything that characterizes the twenty-first century it is that of turbulence. As Kohl and McCool (2016) argue, our context is so volatile, uncertain, complex and ambiguous that the way we plan and manage everything needs to change because, first, the world as we see it has changed, and second, because the basic assumptions we have of the world—that it is Predictable, Linear, Understandable, and Stable (referred to as the "PLUS" world) are no longer valid. Because we act as if we live in a PLUS world, we often fail in our planning and management. And many times, it is not just that we fail, but that our actions will sometimes have the opposite effect of what we intended. Now this may be controversial for many readers, but the trends identified earlier occur at different scales and with varying speeds, making uncertainty and nonlinear change a fact of life.

These assumptions led to a paradigm of decision making that was reductionistic—simplifying what were complex systems of components into something that, perhaps a miscasting, we could understand. However, we now realize those assumptions are no longer appropriate or helpful, because they lead us into framing questions, issues and challenges that do not provide useful insights. Kohl and McCool suggest we view the world as Dynamic, Impossible to completely understand, Complex, and Ever-changing (DICE). By changing our assumptions in this way, we see the world through a new lens—a new paradigm—that leads us to new ways of conducting science, decision making and planning.

For example, for decades, tourism marketing organizations assumed that by generating inquiries resulting from magazine advertisements for more information and by sending fulfillment packages to the inquirers, people would in return be convinced to visit the destination. Research showed, however, that conversion rates were relatively low and could not necessarily be predicted by the promotional content. Such promotional-dominated marketing was soon replaced by internet promotion which transformed an assumed linear decision-making process into a complex web of decisions, competing information sources, visitor priorities and preferences, and competitive pricing strategies all with feedback loops. Initially, internet promotion appealed to a relatively small set of tech-savvy users, primarily the young, active and wealthy, interested in exotic places and adventures. However, as the internet grew in importance, promotion of destinations, while still significant, has been expanded to include wholesalers, custom travel itineraries, non-profit organizations and socially aware providers. This web of interactions, linkages and strategic nodes is itself a complex system. One result is that it has replaced the notion of a value chain with that of a value *network*, making marketing and delivery of recreation opportunities more complex than ever.

The results of promotion are often unexpected, and not always positive. Often conceived as an efficient and benign strategy for generating foreign exchange, citizens of cities in Europe and other places erupted in 2016 and 2017 in protest against rapidly growing visitation in their cities and neighborhoods. This was colloquially termed "overtourism" because of the congestion and unappreciated behavior of visitors. These protests were unexpected, and to some extent ignored by marketing organizations that heavily promoted tourism. This explosive outbreak of protest shows how marketing organizations, dominated by a promotional mental model of tourism development, forgot that marketing includes product development and quality maintenance.

Managers of many national parks and protected areas, and some visitors, likewise express concerns over rapidly increasing visitation. Such visitation is claimed to damage natural and cultural heritage and all but ruin visitor experiences. In 2017, for example, a year following the U.S. National Park Service 100th anniversary and the launch of its app "Find Your Park" designed to boost visitation, newspapers were peppered with headlines such as "Parks Being Loved to Death," "How Many is Too Many?" and so on. These articles expressed concerns not only about rapidly rising use levels but the consequent impacts of visitation on natural heritage, and on the visitor experience as well. In short, the "fix" of increased promotion has failed to provide the higher quality, transformative experiences that the American National Park Service envisioned in 2012 through its Advisory Board (National Park System Advisory Board 2012).

These "surprises" experienced in both cities and parks have come about because of the growing complexity of the social-ecological systems in which we live, the lack of understanding of those systems, and the absence of a vision of what sustainable tourism is all about. They represent an underlying assumption, often only implicit, that more is better.

This complexity is not a figment of our imagination but characterizes all but the most simple and trivial decisions, leading to wicked problems and messy situations. Wicked problems, briefly described, are those in which society lacks agreement on goals and there is a lack of agreement on cause–effect relationships. Rittel and Webber (1973: 160) argue that:

> As distinguished from problems in the natural sciences, which are definable and separable and may have solutions that are findable, the problems of governmental planning—and specially those of social or policy planning—are ill-defined; and they rely upon elusive political judgment for resolution.

Wicked problems have no *solution* (in the sense of an answer) but are *resolved* (by agreement or consensus) through political discourse that often uses science as a source of information. Such agreements/resolutions are temporary because the context for them is in a constant state of change. The issue of tourism development and management is the consummate example of a wicked problem, for goals are often left unspecified or frequently articulated in vague or ambiguous terms, focus on inputs (e.g., what actions to take, or, worse, a proposal for an increase in visitation), or at best lack a sense of agreement among the destination populations most affected. There is only an illusion of understanding of causes and effects.

Systems thinking as a response to complexity

This complexity challenges conventional, reductionistic paradigms of science, including that focused on tourism and recreation. But when "facts are uncertain, values in dispute, stakes high and decisions urgent"—certainly a description of many tourism situations—a different kind of science is needed (Funtowicz and Ravetz 1995). Wicked problems occur in situations where scientists disagree on cause–effect relationships and social agreement on goals does not exist. And because of these situations we need to look carefully at the suitability and usefulness of the paradigms currently used to guide research.

How problems are defined in simple, linear systems is evident, but for wicked problems, the definition is "in the eye of the beholder" (Allen and Gould 1986). In wicked systems, problems are defined through public engagement, deliberation, reflection and debate. They are resolved only through similar processes, and therefore the important test of "answers" is their usefulness, rather than their "correctness," which really cannot be determined in messy situations. The way forward in complex situations is to better understand the character of the problem through public engagement and systems thinking. This argument is articulated in this extensive quote from Morris and Martin (2009: 156):

> The journey towards sustainability is a 'wicked' problem involving complexity, uncertainty, multiple stakeholders and perspectives, competing values, lack of end points and ambiguous terminology . . . In a word, dealing with sustainability means dealing with a mess and most people avoid messes because they feel ill-equipped to cope. The health, agricultural, financial and ecological problems we now face are qualitatively different from the problems for which existing scientific, economic, medical and political tools and educational programmes were designed. Without the right tools, learners faced with these wicked problems may fall back on the same old inappropriate toolbox with at best, disappointing outcomes. Given the messy nature of the dilemmas and contradictions facing us there can be no single recipe and no definitive set of tools. Yet some ways of thinking and of doing things do seem more useful than others in this context. These approaches are as much about 'problem finding' and 'problem exploring' as they are about problem solving. Our contention is that learners cannot deal with the wicked problems of sustainability without learning to think and act systemically.

Sustainable tourism, I believe Morris and Martin would agree, falls within this argument. It is not the simplistic finding of the intersection of ecological viability, economic feasibility and social acceptability. Such an approach implicitly assumes a stable balance among these three partly competing, partly overlapping values. But the world is hardly stable and such assumed balances soon become undone, if they ever were in balance, and thus upset plans long made with this balance implicit. The world is in a constant state of change; some changes are local and fast; others are global and likely slow.

The systems-thinking viewpoint is becoming more popular among tourism scientists (e.g., Farrell and Twining-Ward 2004; Strickland-Munroe, Allison and Moore 2010; Bricker et al. 2015; Hughes, Weaver and Pforr 2015; Kohl and McCool 2016; McCool and Bosak 2016; Espiner, Orchiston and Higham 2017) as a way of dealing with complexity and complex systems. So systems thinking is not just a new paradigm; it is a tool to build not only the situational awareness needed in dynamic situations (see e.g., Nkhata and McCool 2012) but it helps build foundations for developing resolutions to problems.

Systems thinking, according to Senge (2006: 68–9),

> is a discipline for seeing wholes. It is a framework for seeing interrelationships rather than things, for seeing patterns of change rather than static 'snapshots' . . . Today systems thinking is needed more than ever because we are becoming overwhelmed by complexity.

The focus in systems thinking is on the connectedness of actors/actions rather than the characteristics of actors/actions. Systems thinking assumes nonlinear causality and displays connections as a series of feedback loops rather than assuming linear uni-directional connections between causes and effects. Rather than assuming that a one-time event underlies consequences, systems thinking views causality as an ongoing process, with consequences providing feedback to causes. Tourism research has been dominated by conventional thinking, focused on developing knowledge about the parts of a system for the last 50 years. And while this focus has led to an increase in knowledge, *we are now at a point where we need to focus on the system rather than its parts to deal with the complexity, change, and uncertainty that characterizes the context for tourism. We need to focus on developing understanding rather than simply knowledge.*

Systems thinking starts with defining a distinctive whole and then examining the interconnected parts or components. Systems thinking recognizes that the whole contains *emergent* properties, that is, characteristics of the whole that cannot be predicted from its parts, just as the thinking property of the human brain cannot be predicted by examining its neurons or the transportation property of an automobile could not be predicted by finding a steering wheel in the desert. A system in this context includes both social and ecological components as both are important components of much of tourism.

In Stage I, we define a system of interest and its boundaries (we note that defining boundaries is not easy, and ultimately there are no exogenous forces or processes). In Stage II, we simplify the system to better understand it and use our knowledge to resolve problems. Stage II represents an effort to build a model of the system—a model being a simplistic representation of the system. This model building and simplification occurs only *after* we understand the complexity of the situation. Simplifying the system only comes after we have considered the system's distinctive whole and its interconnected parts. The purpose of the modeling is more in the way of understanding how the system works rather than prediction. Stage II helps us understand where delays occur, where leverage points in changing the system outputs arise and how we might organize to engage the complexity of the system. Inevitably, our models will be wrong (Sterman 2002) which means we think and act adaptively to change in response to change.

Table 1.1 Best practices for functioning in a complex situation

Complexity practice	Short description
Build situational awareness	Heightened awareness to sense the unexpected
Invest in personal relationships	Engaging in collaborative forms of governance needed to manage complex situations
Appreciate the power of networks	Networks are needed to characterize simply and engage complexity
Seek and use leverage points	Leverage points are places in a system to make change
Employ different forms of knowledge	Different forms of knowledge help us understand complex systems
Learn continuously	Uncertainty pervades our world; what once worked may no longer work

Note: Following McCool et al. (2015). (For a complete discussion of these practices see that volume.)

In Stage III, we engage this complexity, and do so by implementing in various forms complexity practices identified by McCool et al. (2015). These practices (Table 1.1) are implemented based on an understanding of the adaptive character of the system, what the system whole may be, and the model we have developed. For example, Anderies, Janssen and Ostrom (2004) have suggested a model of natural resource systems that is scale independent and emphasizes relationships among providers and regulators of interactions in these systems. The process of implementing the practices in Table 1.1 build the social capital needed to address the problems of complex, dynamic systems.

In Figure 1.2, we show a simple depiction of how overtourism can be viewed. Visitation rises and leads to impacts, social and biophysical. A standard of how much change is acceptable exists (however, often only implicit) leading to a gap between what exists and what is acceptable. This leads to management action, often called establishing a carrying capacity implemented through a use limit. The system assumes impacts are linearly related to the number of visitors, and thus, impacts are reduced from implementing the use limit. Unfortunately, this system is simplistically represented, as are the fundamental assumptions at the foundation of carrying capacity (see McCool and Lime 2001 for a critique). Use limits often lead to surprises, after some kind of delay. Surprises are the unanticipated consequences. For example, implementation of a use limit may shift the burden (Senge 2006) of resolving impact issues to other visitors, other managers (as visitors shift to other areas), or residents of other communities. This systems-thinking approach helps us better understand what happens when we don't perceive the underlying system.

Systems thinking helps us understand not only what is happening in a complex system, but also how the parts are related. This increased understanding helps us not only to propose actions that impact the functioning of the system as a whole,

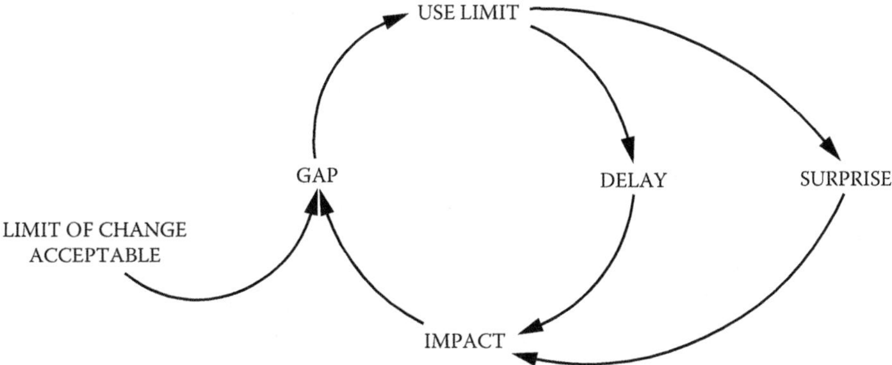

Notes: Here the system "begins" with a level of impact on social systems; this impact is compared, either explicitly or implicitly, to some idea of whether the impact is acceptable, and this leads to selection of a management strategy, here noted as a use limit. This limit leads to, after a delay, a reduction of impact, which in turn will be compared to the standard of acceptability. Sometimes, as shown here, a "surprise" or unanticipated consequence occurs, leading to more or less impact, and the result feeds through the causal loop again.

Figure 1.2 Simple depiction of overtourism using a causal loop diagram

but also may help the system achieve better *resilience* in the face of the global level processes affecting it. Resilience is the ability of a system to respond to disturbance and retain its structure and function. Resilience is not only a characteristic of the whole system—it can become a goal for management. And by focusing on resilience as a goal, we then also tend to the system as a whole.

For example, we can ask how tourism can be managed instead of or in addition to use limits to make a social-ecological system more resilient, and potentially less vulnerable, to the global-level stresses and strains we all face. We can ask the question, *what is it that tourism sustains and should sustain,* rather than seeking some mythical conjunction of social acceptability, economic viability and ecological responsibility, all of which represent value judgments on the part of various publics and interests.

Conclusion

This book is all about identifying the information challenges for sustainable tourism in a century of turbulence as well as opportunity—the turbulence we experience every day as we go to the office or visit the field; the opportunity provided by abandoning the paradigms of the past (unlearning, if you will) that worked well once but now handicap our thinking, and advancing new paradigms which provide a more useful lens through which we see opportunity where others have only seen problems. It is about providing a base for producing policy-relevant research, research that informs management or policy making when it comes to choosing among alternatives, understanding the character of a wicked problem, or even framing a messy situation.

Understanding the challenges we have before us through building our awareness helps enhance system resilience. We have included chapters on many, but not all, of the various dimensions or arenas of sustainable tourism, the content of which is influenced by the question of what it is that tourism should sustain. The topics in this book are not necessarily limited to one kind of situation, type of state, geography or constituency. Each one is applicable to all kinds of contexts, but of course are modified to be applicable to any particular social, political or geographical environment. They are designed to stimulate research, and to provide information and knowledge regardless of where one lives, who a person or community may be, or what one does in the context of tourism.

In the following chapters, various authors describe what they believe to be the important research challenges and opportunities by identifying information gaps, needs and significant questions for research, with the result that our understanding of how various tourism-related systems function and how they can be better managed to be resilient in the face of change is enhanced.

Together, they form an *agenda for research* to make tourism more sustainable, a great need in a world of increasing complexity and scarcity. They suggest that perhaps some of the research questions we as a field have pursued in the past may no longer need to have our attention. They provide opportunities to lay a *foundation* for addressing the wicked and messy situations of the future. We discuss them now because with substantial tourism growth projections for the near future, we need "to get our act together" to avoid unnecessary turmoil and conflict.

This is particularly important when we view tourism as a developmental action, forcing us to address the questions of what tourism sustains, how we can make tourism development more efficient, effective and equitable, and how we can improve tourism research to make it more relevant and useful in addressing these issues.

References

Allen, G. M. and Gould Jr, E. M. (1986). "Complexity, wickedness, and public forests." *Journal of Forestry*, **84**(4): 20–23.

Anderies, J. M., Janssen, M. A. and Ostrom, E. (2004). "A framework to analyze the robustness of social-ecological systems from an institutional perspective." *Ecology and Society*, **9**(1): 18. Available at http://www.ecologyandsociety.org/vol9/iss1/art18/.

Bricker, K. S., Donohoe, H., Becerra, L. and Nickerson, N. (2015). "Theoretical perspectives on tourism—an introduction." In K. S. Bricker and H. Donohoe, eds., *Demystifying Theories in Tourism Research*, pp. 1–6. Wallingford: CABI.

Drumm, A. (2007). "Tourism-based revenue generation for conservation." In R. Bushell and P. Eagles, eds., *Tourism and Protected Areas: Benefits Beyond Boundaries*, pp. 191–209. Wallingford: CABI.

Espiner, S., Orchiston, C. and Higham, J. (2017). "Resilience and sustainability: a complementary relationship? Towards a practical conceptual model for the sustainability–resilience nexus in tourism." *Journal of Sustainable Tourism*, **25**(10): 1385–1400.

Farrell, B. H. and Twining-Ward, L. (2004). "Reconceptualizing tourism." *Annals of Tourism Research*, **31**(2): 274–95.

Funtowicz, S. O. and Ravetz, J. R. (1995). "Science for the post normal age." In L. Westra and J. Lemons, eds., *Perspectives on Ecological Integrity*, pp. 146–61. Dordrecht: Springer.

Hughes, M., Weaver, D. and Pforr, C., eds. (2015). *The Practice of Sustainable Tourism: Resolving the Paradox*. London: Routledge.

IUCN (International Union for the Conservation of Nature) (2014). "A strategy of innovative approaches and recommendations to improve health and well-being in the next decade." Gland: IUCN.

Kohl, J. and McCool, S. F. (2016). *The Future has Other Plans*. Denver, CO: Fulcrum.

McCool, S. F. and Bosak, K., eds. (2016). *Reframing Sustainable Tourism*. Dordrecht: Springer.

McCool, S. F., Freimund, W. A., Breen, C., Gorricho, J., Kohl, J. and Biggs, H. (2015). "Benefiting from complexity thinking." In G. L. Worboys et al., eds., *Protected Area Governance and Management*, pp. 291–326. Canberra: ANU Press.

McCool, S. F. and Lime, D. W. (2001). "Tourism carrying capacity: tempting fantasy or useful reality?" *Journal of Sustainable Tourism*, **9**(5): 372–88.

Morris, D. and Martin, S. (2009). "Complexity, systems thinking and practice." In Arran Stibbe, ed., *The Handbook of Sustainability Literacy*, pp. 156–64. Totnes: Green Books.

National Park System Advisory Board (2012). "Revisiting Leopold: resource stewardship in the national parks." Washington, D.C.: National Park System Advisory Board, National Park Service.

Nkhata, B. A. and McCool, S. F. (2012). "Coupling protected area governance and management through planning." *Journal of Environmental Policy & Planning*, **14**(4): 394–410.

Rittel, H. W. and Webber, M. M. (1973). "Dilemmas in a general theory of planning." *Policy Sciences*, **4**(2): 155–69.

Senge, P. M. (2006). *The Fifth Discipline: The Art and Practice of the Learning Organization*. London: Random House.

Shakya, M. (2016). "Tourism and social capital: case studies from Nepal." In S. F. McCool and K. Bosak, eds., *Reframing Sustainable Tourism*, pp. 217–39. New York: Springer.

Sterman, J. D. (2002). "All models are wrong: reflections on becoming a systems scientist." *System Dynamics Review*, **18**(4): 501–31.

Strickland-Munro, J. K., Allison, H. E. and Moore, S. A. (2010). "Using resilience concepts to investigate the impacts of protected area tourism on communities." *Annals of Tourism Research*, **37**(2): 499–519.

United Nations (2015). "Transforming our world: the 2030 agenda for sustainable development." New York: United Nations.

UNWTO (UN World Tourism Organization) (2005). "Making tourism more sustainable: a guide for policy makers." Madrid: United Nations.

UNWTO (UN World Tourism Organization) (2011). "Tourism towards 2030/global overview." Madrid: United Nations.

UNWTO (UN World Tourism Organization) (2017). "2016 Annual Report." Madrid: United Nations.

World Travel and Tourism Council (2017). "Travel and tourism economic impact 2017." London: World Travel and Tourism Council.

2 The tourism system

Keith Bosak

Introduction

One of the fundamental challenges facing tourism research is that tourism itself has been difficult to define and thus differentiate from other activities. Is tourism an industry or multiple industries? How do researchers sort out the multiple overlaps between tourism and other economic activities? Given that tourism is difficult to define and differentiate, how do we as researchers understand its impacts? These are questions that often arise in tourism research and have been partially addressed by conceptualizing tourism as a system. However, even conceptualizing tourism as a system has had its challenges. One of the primary questions has been: What are the components of the system and how do they interact? This question has been addressed in the literature on tourism since the 1960s and various authors have characterized the tourism system using a variety of models (Pearce 1995). Today, tourism researchers are also dealing with these questions in the face of rapid social and environmental change and in a world that is increasingly complex and interconnected. The previous models of the tourism system have played an important role in developing our understanding but there is a need to rethink these models in the context of today's (and tomorrow's) world.

In this chapter I address the challenge of rethinking the tourism system first by providing a characterization of the tourism system as it has been conceptualized in the literature. We then provide a synthesis of the various descriptions of the tourism system paying attention to common characteristics. Ultimately, this synthesis leads to a deeper understanding of the evolution of conceptualizations of the tourism system. This understanding is particularly salient in the context of today's complex and rapidly changing world where (sustainable) tourism is increasingly seen as a path to development that also protects the environment.

Given the difficulties in understanding tourism, its components and interactions, researchers increasingly recognize the need to employ a Socio-Ecological Systems (SES) approach to conceptualizing the tourism system to address the complexity found in today's world. Conceptualizing tourism as an SES addresses many of the shortcomings of prior conceptualizations and allows researchers to view tourism as a self-organizing system with emergent properties that occurs at multiple scales, has impacts at multiple scales and is part of various feedback loops. I conclude with a discussion of the implications for research in conceptualizing through an SES framework.

Characterizing the tourism system

Tourism researchers have characterized the tourism system in a variety of ways. Early on, researchers attempted to develop spatial models for tourism and tourism development (Pearce 1995). These models tended to focus on the spatial interactions that characterize tourism and their associated patterns and processes. Pearce (1995: 3) identified four basic groups of models originating from geographic inquiry: Linkage or travel-focused, origin and destination, structural, and evolutionary. Although each type of model focused on different aspects of tourism, most were founded on the origin-linkage-destination concept to explain tourism patterns and processes (Pearce 1995). These models were useful in conceptualizing tourism as a spatial phenomenon as well as an economic activity. However, most did not adequately address the tremendous complexity of the tourism system and instead tended to focus on basic concepts.

The 1970s and 1980s represent a second era in the development of tourism system conceptual models. A number of notable and important models came from this era. These include models by Gunn (1979), Leiper (1979), Mathieson and Wall (1982), and Mill and Morrison (1985). Gunn proposed a system consisting of a demand side and a supply side. The demand side consists of the population with an interest and ability to travel. The supply side centers on the attraction and also includes services, information/promotion and transportation. The dynamic is one where the attraction(s) promote and provide information to the population who then use transportation to visit the attraction and utilize the services (Gunn 1979). Gunn recognized that the system was dynamic and influenced by external factors that can impact its functioning. Mill and Morrison (1985) also conceptualized the tourism system in a similar fashion. Their model contained four components: market, travel, destination and marketing. However, this model considers the interaction of these components using a consumer behavior approach, answering the questions of where, when and how to travel once the decision to travel has been made. Although the tourist and their behavior are at the forefront of this model, it functions similar to Gunn's in that the destination develops marketing to reach the market and encourage them to travel. Those people then decide to visit the destination and therefore begin to make choices about the transportation and services they will use.

Mathieson and Wall (1982) took a different approach to conceptualizing the tourism system. Their approach involved three elements (dynamic, static and consequential). The dynamic element focuses on the travel to a destination. The static element is the stay at the destination and the consequential element is the outcome of the first two elements. The consequential element includes the physical, social and economic subsystems the tourist(s) impacts. This model focuses on impacts and is designed as such. The dynamic and static elements combine in myriad ways to produce an array of diverse impacts. This model differs from the previous two in that it does not separate the tourism system into supply and demand sides but rather focuses on the processes and variables that interact to produce impacts. Leiper also moved conceptualizations of the tourism system forward by explicitly using a systems methodology and envisioning five elements (tourists, generating region, transit route, destination region and tourist industry). Leiper's (1979: 404) model focuses on spatial and functional connections and acknowledges but does not specifically situate the broader environments within which the system operates (physical, cultural, social, economic, political and technological). This model was designed to capture the geographical, behavioral, industrial and environmental elements of the system in a broad sense.

The models developed during this era can be characterized as focused on the flow of tourists to and from destinations with impacts occurring during their time traveling. These models acknowledge external factors and complexity but do not explicitly address either. Leiper, for example, adds the environment to his model as does Gunn but neither explicitly address the interactions between the environment and the system. The environment is static and treated as an external factor. Mathieson and Wall acknowledge impacts but only in one direction (the tourist on the destination) and thus their model ignores feedback mechanisms. None of the models adequately address the issue of scale. Even with their limitations, the models developed in the 1970s and 1980s have proven to be quite useful in understanding the basic functioning of the tourism system and are still being used today by students, faculty and researchers in tourism. However, by the 1990s researchers were beginning to see the need to develop models to better engage with the complexity of the tourism system, particularly in light of the recent emergence of sustainable development and sustainable tourism.

In 1995, the *Journal of Sustainable Tourism* published a conversation between Michael Hall and Richard Butler centered on ideas of complexity, sustainability and process within the tourism system. This conversation provides an enlightening glimpse into the state of tourism research at the time and calls for a more holistic, systems-based approach to tourism research that can address complexity, change and scalar interactions. In discussing the current state of tourism research, Michael Hall states:

> Given that a systems approach in tourism has been used as a framework for quite a number of texts, one would have thought that the emphasis on dynamic interrelationships between components of a system would have encouraged tourism researchers to think holistically. However, this has not been the case. I think that this is because the full implications of systems thinking, the role of dynamic and inter-related economic, socio-cultural and environmental factors, and the central importance of process, all of which are critical for understanding sustainability, have not been appreciated by researchers. (p. 101)

Hall points out that researchers had not yet embraced systems thinking, and at the same time outlines the set of factors that need to be addressed in order to pursue a research agenda based on a systems approach. Since this conversation was published, tourism research has taken a slow turn towards a more holistic and systems-based approach.

This turn began with literature that started to address the tourism system as complex and chaotic. McKercher (1999) argued that tourism functions in a non-linerar and non-deterministic fashion and can better be explained as a system using chaos and complexity theory. McKercher notes that previous models acknowledge complexity, but in their reductionist approach, fail to acknowledge the chaotic and non-linear nature of tourism. He also points out that previous models do not capture the "complex array of interactions" (McKercher 1999: 427) and power dynamics that actually occur in tourism. As an alternative to previous models, McKercher proposes a chaos model of tourism that is based on elements of chaos and complexity and treats the tourism community as functioning in many ways like an ecological community. Implicit in this model is also the notion of an adaptive cycle where disturbances cause upheaval and reorganization of the system.

In looking at the interactions between tourism and economic development, Milne and Ateljevic (2001) advocate the embracing of complexity and specifically introduce ideas of scale through a conceptualization of tourism interactions employing the idea of the global–local nexus. They note that relationships between tourism and economic development are non-linear and messy. Ultimately they call for tourism researchers to embrace the complexity in tourism.

Farrell and Twining-Ward (2004) take notions of complexity one step further in their call for tourism researchers to look to complexity theory, ecology, ecological economics and global change science in order to move away from reductionism and towards an integrative approach of natural and human systems. The authors argue that in order to more thoroughly understand tourism, natural and human systems must be thought of as integrated rather than separate. They explicitly invoke the adaptive cycle from Gunderson and Holling (2002) to explain tourism as both complex and adaptive. Ultimately, they develop a Complex Adaptive Tourism System (CATS) model that has been applied heavily in tourism research.

McDonald (2009) links complexity theory and sustainable tourism by taking a complex systems approach to argue that the problems with sustainable tourism development can be better understood "by identifying underlying behaviours influencing a system and providing opportunities to intervene within a system to ensure sustainable outcomes prevail" (p. 469). This article represents a shift towards integrating ideas of sustainability into a complex and adaptive tourism system. Strickland-Munroe et al. (2010) discuss the integration of resilience concepts in investigating impacts of tourism in communities living in and around protected areas. Their argument is that traditional sustainability indicators are insufficient for understanding impacts, and that a resilience approach acknowledges that conditions are rapidly changing and largely unpredictable and therefore require alternative approaches to assessment of impacts. Their article provides an additional link in conceptualizing tourism systems as socio-ecological systems and acknowledging that resilience is a desired condition.

A parallel approach to conceptualizing tourism systems as complex and adaptive also occurred around this same time and can be characterized by statistical modeling approaches that illustrate the non-linearity and unpredictable nature of tourism (Baggio 2008, 2013). Concurrently, work in ecological modeling explicitly addressed socio-ecological systems approaches for sustainability (Petrosillo et al. 2006, Lacitignola et al. 2007, Biggs 2011, Wei, Alvarez and Martin 2013). These studies employ theoretical/conceptual models of resilience and quantitative ecological, social and economic data to develop and test statistical relationships between variables. These approaches attempt to capture complex interactions at specific destinations or regions and their implications for resilience through statistical models. The outcomes are useful in illustrating specific interactions in isolated parts of the socio-ecological system. The body of work in ecological modeling represents the first attempts to address, through empirical analysis, tourism as a socio-ecological system and resilience as a desired outcome.

Cochrane (2010) explicitly links resilience, tourism and sustainability through her conceptualization of the "sphere of tourism resilience" where the triple bottom line of tourism sustainability is combined with elements of a resilient tourism system. The contribution of this article in going beyond other thinking on socio-ecological systems and resilience in tourism is that she argues that the tourism system's unpredictability and multiple variables mean that quantitative measurement of system indicators is likely to be insufficient in producing practical policy outcomes. This "means accepting the predominant validity of the interpretivist paradigm in real-world research" (Cochrane 2010: 183) in order to escape the temptation of reduction and simplification. Luthe and Wyss (2014) in their research propose the linking of quantitative and qualitative data to further understand resilience. They propose that this be accomplished through quantitative network governance approaches that are then backed up with qualitative research to validate assumptions about the system.

Other recent literature attempts to move beyond current socio-ecological systems conversations in addressing resilience in tourism systems. Ruiz-Ballesteros (2011) developed an approach to looking at the role of community-based tourism in the resilience of the socio-ecological system of Agua Blanca, Ecuador. This study utilized qualitative and quantitative data collected over a three-year period in the community of Agua Blanca to determine if community-based tourism was contributing to resilience based on four criteria derived from the resilience literature. Lew (2014) proposes a Scale, Change and Resilience (SCR) that applies resilience thinking in terms of spatial and temporal scales and in terms of how a system responds to change. Cheer et al. (2017) apply the SCR in the context of islandscapes to show vulnerability and adaptation to tourism disturbances. The development of the SCR signals a change in thinking in tourism research, one that employs an SES perspective and is applicable in addressing real-world challenges. McCool and

Bosak (2016) also move beyond current thinking on SES and sustainable tourism by arguing that a fundamental reframing of the tourism system is needed as the traditional three-legged stool approach is simplistic and reductionist. They argue that resilience is a better way to frame sustainability, and that by reframing sustainable tourism using SES it is possible to address the complex issues and wicked problems associated with tourism in general and sustainable tourism specifically.

Conclusion

Models of the tourism system have evolved over time as researchers have attempted to address the complexity inherent in tourism. Early models of the tourism system focused on flows of tourists and money from source region to destination and back with impacts occurring along the way. These approaches were (and still are) useful in gaining a basic understanding of tourism components but they fail to adequately address the complexity and feedbacks of the system, treat the environment as static and can't account for the self-emergent qualities of tourism development. These shortcomings have been incrementally addressed since the 1990s with a call for a more holistic approach to understanding the tourism system. First, complexity and chaos theories were invoked as explanation, leading to the inclusion of resilience and socio-ecological systems approaches. This evolution continues today and will bring about many different approaches to conducting research that will require the designing of new and innovative methods for gathering data to answer questions relating to sustainability in tourism.

These approaches will need to be inclusive of ecological and social components and focus on the interactions between and among various components with an eye to the feedbacks and emergent properties that stem from disturbances with an acknowledgment of the nested properties of systems. Ultimately, an understanding of the tourism system will require novel methods to incorporate quantitative, qualitative and spatial data while employing a paradigm that allows for the co-production of knowledge between researchers and communities. This data will be required to allow researchers to capture the relationships between a dynamic and changing environment (ecological and social), economies, governance systems, scalar interactions and emergent properties. New approaches to visualizing this data will also need to be developed so that the complex interrelationships between social and ecological systems can be represented to decision-makers and communities alike. Ultimately, the turn towards SES thinking and approaches in tourism will allow for research that is more holistic and can provide the knowledge necessary to make interventions in the system that will enhance resilience and thus sustainability.

References

Baggio, R. (2008). "Symptoms of complexity in a tourism system." *Tourism Analysis*, **13**(1): 1–20.
Baggio, R. (2013). "Studying complex tourism systems: a novel approach based on networks derived from

a time series." XIV April International Academic Conference on Economic and Social Development, Moscow, April 2–5, 2013.

Biggs, D. (2011). "Understanding resilience in a vulnerable industry: the case of reef tourism in Australia." *Ecology & Society*, **16**(1): 30.

Cheer, J., Cole, S., Reeves, K. and Kato, K. (2017). "Tourism and islandscapes—cultural realignment, socio-ecological resilience and change." *Shima*, **11**(1): 40–54.

Cochrane, J. (2010). "The sphere of tourism resilience." *Tourism Recreation Research*, **35**(2): 173–85.

Farrell, B. H. and Twining-Ward, L. (2004). "Reconceptualizing tourism." *Annals of Tourism Research*, **31**(2): 274–95.

Gunderson, L. and Holling, C., eds. (2002). *Panarchy*. Washington, D.C.: Island Press.

Gunn, C. A. (1979). *Tourism Planning*. New York: Taylor & Francis.

Journal of Sustainable Tourism (1995). "In search of common ground: reflections on sustainability, complexity and process in the tourism system—a discussion between C. Michael Hall and Richard W. Butler." *Journal of Sustainable Tourism*, **3**(2): 99–105.

Lacitignola, D., Petrosillo, I., Cataldi, M. and Zurlini, G. (2007). "Modelling socio-ecological tourism-based systems for sustainability." *Ecological Modelling*, **206**(1–2): 191–204.

Leiper, N. (1979). "The framework of tourism: towards a definition of tourism, tourist and the tourism industry." *Annals of Tourism Research*, **6**(4): 390–407.

Lew, A. (2014). "Scale, change and resilience in community tourism planning." *Tourism Geographies*, **16**(1): 14–22.

Luthe, T. and Wyss, R. (2014). "Assessing and planning resilience in tourism." *Tourism Management*, **44**: 161–3.

McCool, S. and Bosak, K., eds. (2016). *Reframing Sustainable Tourism*. New York: Springer.

McDonald, J. R. (2009). "Complexity science: an alternative world view for understanding sustainable tourism development." *Journal of Sustainable Tourism*, **17**(4): 455–71.

McKercher, B. (1999). "A chaos approach to tourism." *Tourism Management*, **20**(4): 425–34.

Mathieson, A. and Wall, G. (1982). *Tourism: Economic, Physical and Social Impacts.* New York: Longman Inc.

Mill, R. C. and Morrison, A. M. (1985). *The Tourism System*. Englewood Cliffs, N.J.: Simon and Schuster.

Milne, S. and Ateljevic, I. (2001). "Tourism, economic development and the global–local nexus: theory embracing complexity." *Tourism Geographies*, **3**(4): 369–93.

Pearce, D. (1995). *Tourism Today: A Geographical Analysis*. New York: John Wiley and Sons.

Petrosillo, I., Zurilini, G., Grato, E. and Zaccarelli, N. (2006). "Indicating fragility of socio-ecological tourism-based systems." *Ecological Indicators*, **6**(1): 104–13.

Ruiz-Ballesteros, E. (2011). "Socio-ecological resilience and community-based tourism: an approach from Agua Blanca, Ecuador." *Tourism Management*, **32**(3): 655–66.

Strickland-Munroe, J. K., Allison, H. E. and Moore, S. A. (2010). "Using resilience concepts to investigate the impacts of protected area tourism on communities." *Annals of Tourism Research*, **37**(2): 499–519.

Wei, W., Alvarez, I. and Martin, S. (2013). "Sustainability analysis: viability concepts to consider transient and asymptotical dynamics in socio-ecological tourism-based systems." *Ecological Modelling*, **251**: 103–13.

3 Superseding sustainability: conceptualising sustainability and resilience in response to the new challenges of tourism development

Stephen Espiner, James Higham and Caroline Orchiston

Introduction: the sustainability paradigm in tourism

The theoretical concept of sustainability has much intuitive appeal in the study of tourism. Indeed sustainability became deeply embedded in tourism studies in the years since the UN Brundtland Report was released in the 1980s (World Commission on Environment and Development, 1987). It is now described as "perhaps the most prominent feature of contemporary tourism discourse" (Higgins-Desbiolles, 2010, p. 116). The importance of sustainability emerged initially in the 1980s from growing concern for the predominantly local impacts of tourism (Hall & Page, 1999; Mathieson & Wall, 1982). In challenging the ideology, discourse and hegemony of the UNWTO (Mowforth & Munt, 2008), which promoted tourism as a 'smokeless industry', the attention of the academic community has become increasingly focused on critical questions of tourism sustainability (Edington & Edington, 1986). The sustainability tradition has been taken up in a range of disciplinary fields including social anthropology (MacCannell, 1973), sociology (Cohen, 1972), geography (Duffus & Dearden, 1990), and ecology (Newsome, Moore & Dowling, 2012), which have offered critical insights into questions of local sustainability (Espiner, Orchiston & Higham, 2017). More recently the United Nations Sustainable Development Goals (SDGs) have drawn attention to the broader challenges embodied in tourism within the broader context of sustainable development.

The concept of sustainable tourism has been subject to many years of intense academic critique (Butler, 1990, 1991, 2015; Hall, 1994; Wheeller, 1991, 1995), taken up by stakeholders with specific interests in resource management, environmental conservation, community development and poverty alleviation, and has been applied to individual tourism businesses, destination marketing organisations, local visitor attractions and entire destinations (Espiner, Orchiston & Higham, 2017; McCool et al., 2013). Less dedicated attention has been paid to the ways sustainable tourism can contribute to sustainable development, a diverse set of imperatives that are embodied in the SDGs.

A critical moment in the problematisation of sustainable tourism occurred in the early 2000s when fundamental questions of spatial and temporal scale were raised by authors including Hall (2007b) and Becken and Schellhorn (2007). McCool et al. (2013, p. 217) explain that models of sustainability in the late twentieth century were based on the assumption that the world was ". . . predictable, linear, ultimately understandable and basically stable". This notion has been deeply questioned (Farrell & Twining-Ward, 2005) and it has been argued that sustainability should at the very least be reconceptualised as a dynamic concept (Espiner, Orchiston & Higham, 2017). Indeed Bostrom (2013) distinguishes between the concepts of 'sustainable trajectory' and 'sustainable state'. A sustainable trajectory requires a reconsideration of the spatial and temporal parameters of sustainable tourism (Spector, Higham & Doering, 2017) and a shift away from the prevailing short-term focus on sustainable state (Becken & Schellhorn, 2007; Hall, 2007a; Spector, Higham & Doering, 2017). Thus the assumptions underpinning the concept of sustainability have been challenged by questions of system complexity (McCool, 2015), global environmental risks (Young, Higham & Reis, 2014) and natural disasters (Faulkner, 2001).

The challenge of system complexity in tourism development is highlighted by the global climate crisis (McBoyle & Wall, 1987; Wall, 1993; Wall et al., 1986), which has given urgency to the need to consider new and extended spatial and temporal trajectories in the field of tourism studies (Hall, 2007b). Climate change has challenged the very concept of sustainability (Bostrom, 2013; Karlsson, 2015). Global catastrophic risks (Bostrom & Ćirković, 2008) that threaten irreversable system shifts (Orchiston & Higham, 2016), are fundamentally irreconcilable with local, steady state sustainable tourism. At a different scale, so too are local/regional or national catastrophic natural disasters, such as regionally destructive earthquakes. It is apparent that greater attention must be paid to the complex socio-ecological systems (SES) within which sustainable tourism occurs (McCool, 2015). The emergence of resilience in the tourism discourse, and its relationship with sustainability, may offer insights into achieving effective longer-term planning in tourism studies.

In this chapter, our discussion focuses upon sustainable tourism and resilient communities, and we interrogate the adequacy of sustainability with reflection upon whether the most sustainable destinations are necessarily the most resilient. We also ask if 'resilience' may usurp sustainability in the discourse of tourism business enterprise; or if resilience can be accommodated as an increasingly important part of the sustainability discourse. The chapter is framed within sustainable regional livelihoods, which includes but in fact extends beyond individual tourism businesses, to the wider communities within which those businesses are situated. Within this context, tourism businesses face a range of internal and external challenges, but they are also affected by larger natural (and human) forces that are largely or entirely beyond their control. In recent years, significant environmental change and powerful geological processes have greatly challenged sustainability and resilience-thinking in tourism-dependent communities in New Zealand. In order to advance a conceptual discussion about the relationship between sustainability and resilience, we present several New Zealand case studies, before reflecting on resilience as a new goal for sustainable tourism. The chapter concludes after outlining a possible research agenda for realising resilience in sustainable tourism.

The emergence of resilience in the tourism discourse

Where the sustainability paradigm focuses on the maintenance of a stable world, many scholars are now coming to terms with chaos and unpredictability (e.g. Faulkner, 2001) in social and ecological systems. This shift in thinking is captured in the concept of 'resilience', which addresses the capacity of a system to absorb disturbance and reorganise while undergoing change (Folke et al., 2010; Walker et al., 2004). Over time, the application of the concept has broadened to include linked, non-linear social-ecological systems (SES) providing a theoretical underpinning for addressing unstable and chaotic systems (Becken, 2013; Bosak, 2016; Cochrane, 2010; Lew, 2014; Russell & Faulkner, 2004; Strickland-Munro, Allison & Moore, 2010). Resilience requires adaptive alternatives to address threats from a wide spectrum of natural or human-induced crises and uncertainties, including those precipitated by infrequent, sudden events such as earthquakes; more frequent, yet unpredictable events such as floods; and a range of longer-term incremental hazards associated with climate change. However, while resilience has been considered an attribute that mitigates uncertainty and unpredictability through the adoption of adaptive social and business practices (Espiner, Orchiston and Higham, 2017), change must be considered through the SES lens (McCool, Freimund & Breen, 2015). Insufficient acknowledgement of deep complexity can lead to unexpected outcomes, resulting in "... more problems, less resilience and to developmental trajectories that are more challenging to deal with" (McCool, Freimund & Breen, 2015, p. 296).

Due to the complexity of tourism systems and their inherent vulnerability to external threats such as natural hazards, or social, political and economic crises, chaos can rapidly envelop tourism activities (Faulkner, 2001). Recognising this, resilience has been an increasing focus of academic endeavour in tourism studies, including conceptual discussions (Bec, McLennan & Moyle, 2015; Bosak, 2016; Cochrane, 2010; Espiner, Orchiston & Higham, 2017; Lew, 2014; McKercher, 1999; Orchiston & Espiner, 2017) and application to tourism case studies (Bec, 2016; Becken, 2013; Biggs, 2011; Espiner & Becken, 2014; Farrell & Twining-Ward, 2004; Orchiston, 2013). There is now considerable consensus that resilience thinking is valuable in addressing uncertainty, where a full understanding of the existing and potential future riskscape is unknown. Tourism destinations need novel strategies to cope with change and, conceptually at least, resilience offers a useful framework to develop new ways of planning and operating in times of high uncertainty (Luthe & Wyss, 2014). The capacity of tourism enterprises to be agile and adaptive in responding to rapid, unexpected change is one clear point of difference between the concepts of resilience and sustainability (Espiner, Orchiston & Higham, 2017).

Re-thinking conceptual relationships in tourism

From the preceding discussion it is possible to see parallels between the concepts of sustainability and resilience as they apply to tourism, especially as both have been used to interpret components of social, economic and environmental maintenance in destination communities. Despite their coexistence in the academic tourism discourse, the overlap or articulation between the two concepts has attracted limited scholarly attention to date. Exceptions include McCool (2015, p. 233) who, discussing the tourism economy, notes that sustainable tourism is "... not a type or scale of business, rather it is a strategy to build or maintain system resilience". This is part of McCool's critique that twenty-first-century tourism planners and developers should look more earnestly at what tourism can sustain in a community, rather than how tourism activity itself can be sustained. Hence, he argues that "[t]he principal question facing tourism in the 21st century is the extent to which it can contribute to the resilience of communities in this era of integration and globalization" (McCool, 2015, p. 233).

Lew (2014) has also made some conceptual comparisons, developing a model for 'scale, change and resilience in tourism' to suggest that resilience planning may be a more effective development approach than the conventional sustainability paradigm. Moreover, Lew (2014, p. 14) argues that sustainability emphasises mitigation to prevent change, whereas resilience adapts to change by building in capacity "...to return to a desired state following both anticipated or unanticipated disruptions". While this is an important conceptual distinction to make, Lew's distinction does not fully acknowledge the potential of system disruptions to produce new ways of doing things, which may not involve 'return to a desired state'. There is evidence that tourism destination communities can capitalise and build on crisis events, precipitating positive change and more resilient practices—which may, or may not be interpreted as sustainable (Espiner, Orchiston & Higham, 2017).

Given the paradigmatic status of 'sustainability' and the growing importance of 'resilience' in the tourism discourse, it is useful to consider how resilient practices might align with fundamental issues of tourism destination and enterprise sustainability. In an attempt to conceptualise the complex relationships within the tourism system and between the concepts of sustainability and resilience in particular, Espiner, Orchiston and Higham (2017) propose a model (Figure 3.1) illustrative of the nature-based tourism sector in New Zealand, but with potential application to other tourism settings. This conceptual model situates nature-based tourism destinations at the centre of an SES system, subject to continuous local- and global-scale influences which ultimately shape the future of the destination.

Espiner, Orchiston and Higham (2017) use their model to argue that the most sustainable destinations are those with the highest levels of resilience. Because resilience is necessary but not sufficient for sustainability, it is illustrated in Figure 3.1 as specialised spheres of varying sizes, often intersecting with, yet conceptually separate from, sustainability. Resilience, then, may be seen as a 'buffer' or a

Notes: In response to exposure to various perturbations arising within the macro context, resilience is illustrated as a range of shifting spheres—sometimes neatly intersecting with sustainability principles ('mature resilience') and other times not at all ('emergent resilience').

Source: Espiner, Orchiston and Higham (2017).

Figure 3.1 Conceptual relationship between resilience and sustainability

'lubricant' enabling the mechanisms of sustainability. Without resilience, sustainability cannot be realised.

The model characterises the relationship between sustainability and resilience in tourism as one of three potential states: emergent, developing and mature (Espiner, Orchiston & Higham, 2017). Where destinations lack resilience or where resilience is emergent, the sustainability of the tourism system is most vulnerable to the perturbations associated with the wider socio-political, economic and environmental system (Hopkins & Becken, 2015). Destinations (or specific businesses) where resilience is most developed, covering a comprehensive range of potential exposure scenarios, are likely to have the most effective buffers against key threats to sustainability (Espiner, Orchiston & Higham, 2017).

Tourism sustainability and resilience to natural disasters

The conceptual relationship between resilience and sustainability illustrated in Figure 3.1 may inform a discussion of the nature-based tourism sector in New Zealand that offers insights into the challenges and opportunities of operationalising sustainability and resilience at other destinations that may be centrally, regionally or peripherally located. New Zealand is a geographically distant tourism destination, heavily reliant on inbound international tourism markets, the closest of which requires a minimum of three hours of air travel. At the core of New Zealand's tourism system is a strong nature-based tourism sector. New Zealand's most iconic tourism destinations are located within highly dynamic landscapes and marine ecosystems that have been created by powerful geotectonic forces, and that are subject to abrupt geophysical change (Purdie, Gomez & Espiner, 2015). Across large timescales, floods, earthquakes and volcanic eruptions have shaped environments that now support scenic and adventure-oriented tourism activities—typically based in small, regional communities that have come to depend on tourism to sustain livelihoods, following several decades of sustained high growth in tourism arrivals.

The conceptual model outlined in Figure 3.1 implies that, beyond the immediate spatial (local) and temporal (short-term) dimensions of sustainable tourism, broader resilience planning is required to confront phenomena including climate change, energy scarcity, natural disasters, and economic and political change. In New Zealand natural events such as episodes of extreme weather and flooding, volcanism and destructive seismic events have become critical aspects of sustainable tourism and resilience planning. Damaging earthquakes such as the Wairarapa 1855 magnitude (M) 8.2, Murchison 1929 M7.8, Napier 1931 M7.8, and Inangahua 1968 M7.1 events are deeply embedded in the national consciousness. National awareness of earthquake risk has been re-awakened in recent years due to a period of intense seismic activity centred on a series of active faults associated with the tectonic plate boundary in the middle and upper South Island.

In September 2010, the city of Christchurch, the key aviation gateway to the South Island, was struck by an M7.1 earthquake. Christchurch was significantly damaged; heritage building facades collapsed and the international airport was closed temporarily. Five months later in February 2011 a powerful M6.3 aftershock struck 5km beneath the city. The February aftershock resulted in 185 fatalities and extensive damage to buildings and infrastructure. The aftershocks have continued over subsequent years, and the full social and community impacts of the Christchurch earthquake sequence are not yet fully understood. Eight extremely testing years later, Christchurch remains in the post-disaster recovery phase (Orchiston & Higham, 2016), which has been hampered by continuing seismic activity, slow restoration of infrastructure within the central business district (CBD) which was cordoned off for more than two years, and long and continuing delays in the rebuilding of hotels, a new convention centre, sports stadium and other community facilities and tourism attractions.

In reflecting upon the conceptualization of sustainability and resilience outlined in Figure 3.1, it is important to note that the Christchurch earthquake on 22 February 2011 caused damage that was so extensive that a return to the pre-earthquake tourism status quo has been impossible. Parts of the system were destroyed, particularly the inner city hotel accommodation sector, while other accommodation providers were able to capitalise on their relative lack of damage, location and accessibility. Orchiston, Vargo and Seville (2014) reported that some businesses were significantly better off than others in the aftermath of the quake, which reflected their business agility and resilience.

The Christchurch earthquake sequence has also had significant overflow effects on the wider South Island tourism system. International arrivals into Christchurch have been heavily impacted since February 2011, due to the destruction of tourism supply, infrastructure, attractions and activities. In the interim, the significant reduction in aviation capacity into Christchurch has been partly offset by rapid growth in international and domestic visitor arrivals into Queenstown airport (350km to the south-west). Christchurch offers insights into a case of seismic activity that has persisted over many months and years, long-term destination recovery, and the destruction of an urban tourism system. The destination system is undergoing a transformation that will result in a very different destination offering than before. One important consequence of the Christchurch earthquake sequence has been greatly elevated awareness of the importance of resilience planning at the community level, in both central (urban) and regional contexts in New Zealand. Tourism has featured prominently in the next resilience planning environment, given the important of tourism to the Christchurch rebuild, and the wider regional economic context in New Zealand.

The subsequent Kaikōura earthquake (November 2016) offers the contrasts of medium-term status quo resilience planning. On 14 November 2016 at 12.07 a.m., the Canterbury region was struck by a major M7.8 earthquake centred around the rural settlement of Waiau and causing extensive damage on the north-east coast of the South Island, and to the capital city of Wellington (lower North Island). It became known as the Kaikōura earthquake because the small, tourism-reliant coastal town of Kaikōura was cut-off by road and rail due to the strong tremors and heavy landslides in the mountains, and was affected by a tsunami with a maximum wave height of 2.5m with a 7m run-up on parts of the coast (Duputel & Rivera, 2017).

Kaikōura (population 2,080) is a small coastal community that in the late 1980s achieved international recognition when it developed into an iconic whale-watching destination (Simmons, 2014). It is tenuously situated on the coastal platform on the eastern coast of the upper South Island, sitting at the foot of the Kaikōura coastal mountain range, and on the rim of the Kaikōura Canyon which plunges into the depths of the Pacific Ocean just metres off the coastline. On the coat-tails of the success of Whalewatch Kaikōura, a thriving local tourism industry developed based primarily on tourist experiences of a range of abundant marine wildlife species including New Zealand fur seals (*Arctocephalus forsteri*), Dusky dolphins (*Lagenorhynchus obscurus*) and a range of pelagic birds such as Northern and Southern Royal Albatross (*Diomedea epomophora*), Bullers Mollymawk (*Thalassarche bulleri*) and gannets (*Morus serrator*). Kaikōura is also renowned for interpretation of the marine environment based on indigenous Māori cultural values—the success of which has been a driver of the region's Māori cultural renaissance (Simmons, 2014)—and has received international recognition for its environmental record through accreditation from both EarthCheck and Green Globe.

Given its location, Kaikōura is reliant upon the flow of visitors along State Highway 1 (SH1) which runs along the eastern coastline between Picton and Christchurch (Figure 3.2). The Kaikōura earthquake triggered landslides that extensively damaged SH1, the railway main trunk line (which carries tourists on the Coastal Pacific rail experience) and other secondary roads in the region. Dramatic uplift of the coastal platform caused significant damage to food sources on the coastal margin, most particularly the uplift of *paua* beds, which had to be immediately translocated to below the low tide waterline. Kaikōura was isolated for a period of 35 days (Stevenson et al., 2017), at which point it became possible for recovery workers to access the town via a secondary inland route. Initially, approximately 1,200 tourists were stranded in the township, and were cared for by the local community—they were evacuated over a period of several days by air and sea. Damage to the coastal road was so severe that it was initially considered beyond repair.

Kaikōura was cut off from tourists at a time when local tourism and hospitality industry business operators were anticipating the peak summer season (Statistics New Zealand, 2018). The segment of State Highway 1 north of Kaikōura remained closed for a year while repairs were undertaken on the road and rail networks. As a consequence, a detour inland was established via State Highway 63, which significantly altered the flow of visitors, goods and services around the upper South Island. In contrast to the Christchurch earthquakes, very few commercial or residential

Source: GNS.

Figure 3.2 Landslide on State Highway 1

buildings were damaged, with non-structural damage causing the greatest insured losses (Insurance Council of New Zealand, 2017).

The Christchurch and Kaikōura earthquakes vividly demonstrate the fact that tourism management is challenged by SES contexts that are unpredictable, non-linear and unstable (McCool et al., 2013). However, they also provide insights into seismic events that were manifestly different, with severe but contrasting tourism (re)development outcomes. The Christchurch earthquake sequence caused extensive structural damage to residential and commercial buildings, including critical tourism infrastructure (e.g. hotels). Differential physical impacts occurred across the city, with CBD hotels and hospitality operators being severely impacted, compared to lesser damage for low-rise motels in the outer suburbs. As a consequence, there were opportunities for some in the industry to capitalise on disaster outcomes, particularly as recovery workers came into the city causing a major shortage of accommodation (Orchiston & Higham, 2016). The drop in accommodation capacity has still not returned to pre-earthquake levels. The aftershock sequence continued to cause damage in Christchurch for two years, and minor aftershocks are still being felt, which delayed decisions about the reinstatement of tourism infrastructure. A de-marketing programme was undertaken in the six months after

February 2011, and the first campaign designed to attract visitors to return—the 'Road Trip of the South Island' campaign—continued to exacerbate the tourism losses that afflicted the Christchurch and wider Canterbury (provincial) tourism economy (Orchiston & Higham, 2016). The heritage character and urban tourism product of Christchurch was destroyed to the extent that rebuilding with the aim of returning to the pre-earthquake status quo was impossible.

The Kaikōura earthquake, in contrast, caused major devastation of coastal land forms and the marine environment, most notably shaking and landslides in the coastal range, and dramatic uplift of the coastal platform. There was some disruption of local tourism infrastructure, including the uplifting of the Whalewatch Kaikōura marina (Figure 3.3), which had to be rebuilt to enable commercial operations to resume. While structural damage to buildings in Kaikōura was less pronounced than Christchurch, the destruction of transportation links to the north and south completely disabled the local tourism industry. Only when the inland road reopened and the recovery workforce was able to access the district, was it possible for tourism operators to reopen by offering different products and services (such as accommodation and catering to recovery workers). Successful operators adapted quickly and with agility to capitalise on opportunities, as a way to continue to employ staff and protect the local economy in what were highly challenging

Source: GNS.

Figure 3.3 The marina in Kaikōura showing the results of coastal uplift

market conditions. All of the significant aftershocks in Kaikōura occurred within 24 hours of the earthquake, and the total number of aftershocks since then has been in the lower expected range compared to other earthquake sequences (GeoNet, 2017).

In a final example from New Zealand, we turn to the South Island's West Coast, a region even more geographically peripheral, and with more limited transportation infrastructure than its counterpart destinations on the opposite coast. The West Coast is linked by road and rail to the rest of the South Island; however, access is restricted to alpine passes, with narrow, elongated transport corridors increasing the vulnerability of the highway network (Robinson et al., 2015). Increasingly dependent on tourism to replace the defunct or sunset industries of native timber logging and mining, residents of this region are well-acquainted with the challenges that the area's geography and climate present (Espiner & Becken, 2014).

The topographic high of the Southern Alps coupled with the predominantly westerly flow of weather systems results in high precipitation rates (up to 12m/annum), which creates an extremely active geomorphic and fluvial environment (Robinson & Davies, 2013). In addition, the Southern Alps lie on an active tectonic plate interface, represented by the Alpine Fault, a 450km-long fault that has a long record of generating magnitude 8 earthquakes approximately every 300 years, the last of which was in 1717 (Berryman et al., 2012). It is clear that while the geomorphic, climatic and geologic environment of the West Coast has created areas of high scenic value, it is also a setting that has demonstrated a diverse natural hazard portfolio. Although a major earthquake hasn't been recorded in the region for over 50 years, the estimated future seismic risk suggests very high potential for unprecedented damage to West Coast communities, business and infrastructure, which raises important questions relating to sustainability planning and resilience in the context of the tourism industry specifically.

The threat to the sustainability of West Coast tourism communities is not limited to the effects of possible earthquakes. Espiner and Becken (2014) explored dimensions of the tourism industry's vulnerability and evidence of social adaptation at the Fos and Franz Josef Glaciers in Westland National Park on the West Coast. Among the identified threats to the destination's future were the effects of climate on the accessibility of the glacier attractions, the security of affordable fuel, and potential for a range of natural hazard events (including floods and earthquakes). One major issue in recent years has been the closure of walking access to the glacier due to rapid climate-induced glacial retreat and increased risk of calving at the terminus of the glacier. As a consequence, glacier tourism operators have had to change their business practices, leading to sharp increases in both fixed-wing scenic flights and glacier landings by helicopters. While such business adaptations may represent an entrepreneurial response to the problems of limited physical access and changing climate, questions remain about the medium- and long-term sustainability of this approach given predicted future energy costs, 'acceptable' aircraft noise levels and emission regulations.

The Westland Glaciers destination is a good example of 'developing resilience' (see Figure 3.1). The tourism industry here has been sustained over time despite a range of vulnerabilities (such as changing climate, energy availability/cost and natural disasters), in large measure because of various resilience factors, including an aptitude for diversifying products, community participation in planning, and effective disaster response processes (Espiner & Becken, 2014). Hence this destination does exhibit strong 'pockets' of resilience, evidenced by past and current adaptations to risks and challenges facing business operations. Tourism enterprises operating in these environments are accustomed to preparing for and responding to multiple immediate (e.g. floods, road closures, earthquake risk) or slow onset, long-term (e.g. climate change) vulnerabilities (Espiner & Becken, 2014; Orchiston, 2012; Wilson et al., 2014). But the setting does not qualify as a 'mature' state as resilience is not comprehensive or necessarily coordinated, with some businesses very prepared and others poorly equipped to respond to change. Furthermore, the degree to which all adaptation behaviour can be considered 'sustainable' is uncertain. As noted above, the additional fixed-wing aircraft and helicopter flights—an entrepreneurial response to the problem of limited access—increases the destination's vulnerability to increasing energy costs and contributes to a decline in 'natural quiet'—a recognised quality of national park attractions (Department of Conservation, 2001). While perhaps 'resilient', such strategies could be considered maladaptive and out of alignment with wider sustainability principles.

Is resilience the new goal for sustainable tourism?

The examples drawn from Christchurch, Kaikōura and the South Island's West Coast illustrate the complex relationship between resilience and sustainability, and highlight their respective roles within a given tourism system. Sustainability emphasises aspirational goals associated with the careful use of resources and ensuring provision for future generations. The fragility of those resources, which may be 'public' or common in nature (Heenehan et al., 2015; Ostrom, 2001), and the fact that debate may surround the commitment of common resources to tourism development, lies at the heart of the challenges associated with sustainable tourism. By contrast, resilience is pragmatic and inclusive of a range of responses that may or may not align with sustainability principles. Clearly resilience and sustainability have features that are conceptually similar, but are distinct and, largely (but not necessarily) complementary (Espiner, Orchiston & Higham, 2017). Both have been used as lenses to interpret the social, economic and environmental elements of destination communities.

The cases illustrated in this chapter suggest that resilience planning may be a pragmatic way to reimagine aspirations for nature-based tourism destinations. This reflects a wider recognition that communities need to shift the emphasis of tourism planning from maintaining an unchanging state to responding to inevitable change (Farrell & Twining-Ward, 2004). For a sector faced with a range of major sustainability challenges (Hall, Gössling & Scott, 2015), a resilience approach may be the best way to frame tourism planning and development. It affords deliberate efforts to build capacity to respond to the diverse social and environmental vulnerabilities of peripheral nature-based tourism destinations that operate at various scales (Hall, 2007a).

For nature-based tourism destinations to remain viable over time, resilience and the capacity to adapt to changing and unpredictable conditions will be critical (McCool, 2015). Communities, businesses and conservation managers working in nature-based tourism settings need to consider a range of mechanisms through which to respond to varied perturbations affecting the socio-ecological tourism system (Farrell & Twining-Ward, 2004). The most commonly identified among these are fluctuations in regional, national and international economic and political stability (Hall, 2004). Perhaps less recognised, but increasingly urgent, are the drivers of change illustrated in the case studies above, including natural hazards and climate change, which must be addressed in planning for tourism in peripheral destinations. Failure to incorporate resilience measures into sustainable tourism discourse and future planning frameworks, including assessment of sustainable tourism, is to misrepresent the phenomenon of sustainability. In this sense then, perhaps the aim for nature-based tourism in peripheral settings should no longer focus on 'sustainable' tourism, but rather on 'resilient destinations'. This may be a more useful, attainable and relevant goal for many tourism-dependent communities.

Realising resilience in sustainable tourism: outline of a research agenda

In this chapter we have emphasised the conceptual value in identifying the centrality of resilience in sustainable tourism systems, but it is also important to outline themes and questions for a research agenda that could help underpin the future realisation of resilience and ultimately help sustain the benefits of tourism to geographically peripheral communities. With this goal in mind, we challenge future tourism researchers to develop meaningful indicators of resilience, which will assist destinations in responding and adapting to changes in ways that sustain their enterprises, their communities and their environment. Interdisciplinary and longitudinal empirical studies examining tourism business/community preparedness and responses to multiple vulnerabilities will also be required to deepen our understanding of perceptions or interpretations of risks and challenges facing tourism systems.

Furthermore, we think there is potential to progress the theoretical understanding of the relationship between resilience and sustainability in tourism. First and foremost, while retaining a focus on local/short-term sustainability remains important, the tourism management challenge that we highlight in this chapter is to engage the paradigmatic shift to incorporate cross-disciplinary, multi-agency resilience planning. Public seminars, online resources and targeted training can help communities and key tourism actors to take leadership roles in building 'mature' resilience in destination communities.

There is also an opportunity for researchers to investigate the process of destination communities moving from 'emergent resilience' to 'developing' and 'mature' resilience states (see Figure 3.1). An important aspect of tourism and resilience research is the triggers that may foster the paradigmatic shift in thinking that we have explored here theoretically. The Christchurch earthquake sequence, which began in September 2010, has been the driver of intense interest in resilience planning in central and local government, research communities, and in regional towns and urban centres that serve as both places of residence and work, and as tourism destinations. The identification of barriers and constraints in these processes might enable businesses, communities and governments to fast-track or enhance the likelihood of strengthening social and natural systems against threats to their immediate and enduring sustainability.

Conclusion

Sustainable tourism, for decades interpreted as the product of an enduring balance between socio-cultural, economic and environmental demands, is a concept better served by greater acknowledgement of the role played by resilience attributes. While destinations might be resilient without being sustainable, the reverse is not true; destinations cannot be sustainable if they are not also resilient. Resilience is an inherent dimension of sustainable tourism and needs to be recognised as such and incorporated, alongside community resilience, in tourism planning processes (Espiner, Orchiston & Higham, 2017).

Sustainability and resilience are predicated upon fundamentally different world views. Sustainable tourism can imply an absence of change—maintaining the tourism system more or less in its current state over time, while resilience acknowledges complexity, uncertainty and change (McCool, 2015; Strickland-Munro, Allison & Moore, 2010) and implies adaptation over time (Farrell & Twining-Ward, 2004). As such, resilience may be a more appropriate framework through which to conceptualise tourism in a time of global risks and ecological uncertainties (Young et al., 2015).

This chapter highlights the need for sustainable tourism planning to accommodate resilience in order to (re)focus efforts on coping with ever-changing conditions. As such, it must be noted that the fit between sustainability and resilience may be imperfect. Short-term responses to real and perceived vulnerabilities can lead to maladaptive strategies that work against sustainability (Espiner, Orchiston & Higham, 2017; Hopkins, 2014, 2015). However, incorporating elements of resilience into destination planning offers the potential to create a realistic foundation upon which the aspirational principles of long-term sustainability might be built. It would also confirm the value of both concepts as distinct, albeit overlapping, lenses through which the tourism system, and the complex and dynamic relationships between business, community and environment, can be better understood.

The case examples discussed in this chapter highlight some of the local-, regional- and global-scale catastrophic risks that threaten irreversible system shifts, many of which are fundamentally irreconcilable with traditional notions of steady state sustainable tourism. A new era of tourism planning and management might continue to aspire towards 'sustainability', but in practice, the concept is likely to be superseded by resilience as a more pragmatic, agile and systems-oriented concept. Building social-ecological systems that have adaptive capacity is a critical part of securing positive outcomes for tourism destinations, and will help address McCool's (2015) call for a future focused on how tourism can sustain communities and environments, rather than an emphasis on how tourism itself can be sustained.

In part, the case examples presented in this chapter reveal that peripheral tourism destination communities in New Zealand can often capitalise and build upon crisis events, leading to positive change and more developed resilient practices, through which the benefits of tourism can be sustained. To broaden the applicability of these benefits, the field of tourism studies requires a research agenda that can help identify pathways to resilience in order to assist communities, at various spatial scales, in navigating the likely range of environmental, social and political challenges that threaten to compromise or undermine a common future.

References

Bec, A. (2016). 'Harnessing resilience for tourism and resource-based communities'. Unpublished PhD thesis, Southern Cross University, Australia.

Bec, A., McLennan, C. & Moyle, B. (2015). 'Community resilience to long-term tourism decline and rejuvenation: a literature review and conceptual model'. *Current Issues in Tourism*, **19**(5): 431–57.

Becken, S. (2013). 'Developing a framework for assessing resilience of tourism sub-systems to climatic factors'. *Annals of Tourism Research*, **43**(1): 506–28.

Becken, S. & Schellhorn, M. (2007). 'Ecotourism, energy use and the global climate: widening the local perspective'. In J. E. S. Higham, ed., *Critical Issues in Ecotourism: Understanding a Complex Tourism Phenomenon*, pp. 85–101. Oxford: Elsevier.

Berryman, K., Cochran, U., Clark, K., Biasi, G., Langridge, R. & Villamor, P. (2012). 'Major earthquakes occur regularly on isolated plate boundary fault'. *Science*, **336**(6089): 1690–93.

Biggs, D. (2011). 'Understanding resilience in a vulnerable industry: the case of reef tourism in Australia'. *Ecology and Society*, **16**(1): 30.

Bosak, K. (2016). 'Tourism, development, and sustainability'. In S. McCool & K. Bosak, eds, *Reframing Sustainable Tourism*, pp. 33–44. Dordrecht: Springer.

Bostrom, N. (2013). 'Existential risk prevention as global priority'. *Global Policy*, **4**(1): 15–31.

Bostrom, N. & Ćirković, M., eds. (2008). *Global Catastrophic Risks*. Oxford: Oxford University Press.

Butler, R. W. (1990). 'Alternative tourism: pious hope or Trojan horse?' *Journal of Travel Research*, **28**(3): 40–45.

Butler, R. W. (1991). 'Tourism, environment, and sustainable development'. *Environmental Conservation*, **18**(3): 201–9.

Butler, R. (2015). 'Sustainable tourism: the undefinable and unachievable pursued by the unrealistic?' In T. V. Singh, ed., *Challenges in Tourism Research*, pp. 234–40. Bristol: Channel View Publications.

Cochrane, J. (2010). 'The sphere of tourism resilience'. *Tourism Recreation Research*, **35**(2): 173–85.

Cohen, E. (1972). 'Toward a sociology of international tourism'. *Social Research*, **39**(1): 174–82.

Department of Conservation (2001). *Westland Tai Poutini National Park Management Plan 2001–2011*. West Coast Conservancy Management Plan Series No. 3. Hokitika, New Zealand: Department of Conservation.

Duffus, D. A. & Dearden, P. (1990). 'Non-consumptive wildlife oriented recreation. A conceptual framework'. *Biological Conservation*, **53**(3): 213–31.

Duputel, Z. & Rivera, L. (2017). 'Long-period analysis of the 2016 Kaikōura earthquake'. *Physics of the Earth and Planetary Interiors*, **265**: 62–6.

Edington, J. M. & Edington, M. A. (1986). *Ecology, Recreation and Tourism*. Cambridge: Cambridge University Press.

Espiner, S. & Becken, S. (2014). 'Tourist towns on the edge: conceptualising vulnerability and resilience in a protected area tourism system'. *Journal of Sustainable Tourism*, **22**(4): 646–65.

Espiner, S., Orchiston, C. & Higham, J. E. S. (2017). 'Resilience and sustainability: a complementary relationship? Towards a practical conceptual model for the sustainability–resilience nexus in tourism'. *Journal of Sustainable Tourism*, **25**(10): 1385–1400.

Farrell, B. H. & Twining-Ward, L. (2004). 'Reconceptualizing tourism'. *Annals of Tourism Research*, **31**(2): 274–95.

Farrell, B. H. & Twining-Ward, L. (2005). 'Seven steps towards sustainability: tourism in the context of new knowledge'. *Journal of Sustainable Tourism*, **13**(2): 109–22.

Faulkner, B. (2001). 'Towards a framework for tourism disaster management'. *Tourism Management*, **22**(2): 135–47.

Folke, C., Carpenter, S. R., Walker, B., Scheffer, M., Chapin, T. & Rockström, J. (2010). 'Resilience thinking: integrating resilience, adaptability and transformability'. *Ecology and Society*, **15**(4): 20.

GeoNet (2017). *GeoNet News Special Edition—The Kaikōura Earthquake: One Year On*. GNS Science and the Earthquake Commission publication.

Hall, C. M. (1994). 'Ecotourism in Australia, New Zealand and the South Pacific: appropriate tourism or a new form of ecological imperialism?' In E. Cater & G. L. Lowman, eds, *Ecotourism: A Sustainable Option?*, pp. 137–58. Chichester: John Wiley and Sons.

Hall, C. M. (2004). *Tourism: Rethinking the Social Science of Mobility*. London: Prentice Hall.

Hall, C. M. (2007a). 'North-south perspectives on tourism, regional development and peripheral areas'. In D. K. Müller & B. Jansson, eds, *Tourism in Peripheries: Perspectives from the Far North and South*, pp. 19–37. Wallingford: CAB International.

Hall, C. M. (2007b). 'Scaling ecotourism: the role of scale in understanding the impacts of ecotourism'. In J. E. S. Higham, ed., *Critical Issues in Ecotourism: Understanding a Complex Tourism Phenomenon*, pp. 243–55. Oxford: Elsevier.

Hall, C. M., Gössling, S. & Scott, D. (2015). 'Tourism and sustainability: towards a green(er) tourism economy'. In C. M. Hall, S. Gössling & D. Scott, eds, *The Routledge Handbook of Tourism and Sustainability*, pp. 490–515. London: Routledge.

Hall, C. M. & Page, S. J. (1999). *The Geography of Tourism and Recreation: Environment, Place and Space*. London and New York: Routledge.

Heenehan, H., Basurto, X., Bejder, L., Tyne, J., Higham, J. E. S. & Johnston, D. W. (2015). 'Using Ostrom's common-pool resource theory to build toward an integrated ecosystem-based sustainable cetacean tourism system in Hawai'i'. *Journal of Sustainable Tourism*, **23**(4): 536–56.

Higgins-Desbiolles, F. (2010). 'The elusiveness of sustainability in tourism: the culture-ideology of consumerism and its implications'. *Tourism and Hospitality Research*, **10**(2): 116–29.

Hopkins, D. (2014). 'The sustainability of climate change adaptation strategies in New Zealand's ski industry: a range of stakeholder perceptions'. *Journal of Sustainable Tourism*, **22**(1): 107–26.

Hopkins, D. (2015). 'Applying a comprehensive contextual climate change vulnerability framework to New Zealand's tourism industry'. *AMBIO: A Journal of the Human Environment*, **44**(2): 110–20.

Hopkins, D. & Becken, S. (2015). 'Socio-cultural resilience and tourism'. In A. A. Lew, C. M. Hall & A. M. Williams, eds, *The Wiley-Blackwell Companion to Tourism*. Chichester: John Wiley & Sons.

Insurance Council of New Zealand (2017). *ICNZ Annual Review 2017*. Available at https://www.icnz.org.nz/fileadmin/Assets/PDFs/FINAL_ICNZ_AnnualReview2017_Online.pdf.

Karlsson, R. (2015). 'Three metaphors for sustainability in the Anthropocene'. *The Anthropocene Review*, **3**(1): 23–32.

Lew, A. A. (2014). 'Scale, change and resilience in community tourism planning'. *Tourism Geographies*, **16**(1): 14–22.

Luthe, T. & Wyss, R. (2014). 'Assessing and planning resilience in tourism'. *Annals of Tourism Research*, **44**: 161–3.

McBoyle, G. & Wall, G. (1987). 'The impact of CO2-induced warming on downhill skiing in the Laurentians'. *Cahiers de Géographie de Québec*, **31**: 39–50.

MacCannell, D. (1973). 'Staged authenticity: arrangements of social space in tourist settings'. *American Journal of Sociology*, **79**(3): 589–603.

McCool, S. (2015). 'Sustainable tourism: guiding fiction, social trap or path to resilience?' In T. V. Singh, ed., *Challenges in Tourism Research*, pp. 224–34. Bristol: Channel View Publications.

McCool, S., Butler, R., Buckley, R., Weaver, D. & Wheeller, B. (2013). 'Is the concept of sustainability utopian: ideally perfect but impracticable?' *Tourism Recreation Research*, **38**(1): 213–42.

McCool, S., Freimund, W. A. & Breen, C. (2015). 'Benefiting from complexity thinking'. In G. L. Worboys, M. Lockwood, A. Kothari, S. Feary & I. Pulsford, eds, *Protected Area Governance and Management*, pp. 291–326. Canberra: ANU Press.

McKercher, B. (1999). 'A chaos approach to tourism'. *Tourism Management*, **20**(4): 425–34.

Mathieson, A. & Wall, G. (1982). *Tourism: Economic, Physical and Social Impacts*. London: Longman.

Mowforth, M. & Munt, I. (2008). *Tourism and Sustainability: Development, Globalisation and New Tourism in the Third World* (3rd edn). New York: Routledge.

Newsome, D., Moore, S. A. & Dowling, R. K. (2012). *Natural Area Tourism: Ecology, Impacts and Management.* Clevedon: Channel View Publications.

Orchiston, C. (2012). 'Seismic risk scenario planning and sustainable tourism management: Christchurch and the Alpine Fault zone, South Island, New Zealand'. *Journal of Sustainable Tourism*, **20**(1): 59–79.

Orchiston, C. (2013). 'Tourism business preparedness, resilience and disaster planning in a region of high seismic risk: the case of the Southern Alps, New Zealand'. *Current Issues in Tourism*, **16**(5): 477–94.

Orchiston, C. & Espiner, S. (2017). 'Fast and slow resilience in the New Zealand tourism industry'. In A. A. Lew & J. M. Cheer, eds, *Tourism Resilience and Adaptation to Environmental Change: Definitions and Frameworks*, pp. 250–66. Abingdon: Routledge.

Orchiston, C. & Higham, J. E. S. (2016). 'Knowledge management and tourism recovery (de)marketing: the Christchurch earthquakes 2010–2011'. *Current Issues in Tourism*, **19**(1): 64–84.

Orchiston, C., Vargo, J & Seville, E. (2014). 'Regional and sub-sector impacts of the Canterbury earthquake sequence for tourism businesses'. *Australian Journal of Emergency Management*, **29**(4): 32–7.

Ostrom, E. (2001). 'Reformulating the commons'. In J. Burger, E. Ostrom, R. B. Norgaard, D. Policansky & B. D. Goldstein, eds, *Protecting the Commons: A Framework for Resource Management in the Americas*, pp. 17–42. Washington, DC: Island Press.

Purdie, H., Gomez, C. & Espiner, S. (2015). 'Glacier recession and the changing rockfall hazard: implications for glacier tourism'. *New Zealand Geographer*, **71**(3): 189–202.

Robinson, T. R. & Davies. T. (2013). 'Potential geomorphic consequences of a future great (Mw = 8.0+) Alpine Fault earthquake, South Island, New Zealand'. *Natural Hazards and Earth System Sciences*, **13**: 2279–99.

Robinson, T., Davies, T., Wilson, T., Orchiston, C. & Barth, N. (2015). 'Evaluation of coseismic landslide hazard on the proposed Haast–Hollyford Highway, South Island, New Zealand'. *Georisk*, **10**(2): 146–63.

Russell, R. & Faulkner, B. (2004). 'Entrepreneurship, chaos and the tourism area lifecycle'. *Annals of Tourism Research*, **31**(3): 556–79.

Simmons, D. G. (2014). 'Kaikōura (New Zealand): the concurrence of Māori values, governance and economic need'. In J. E. S. Higham, L. Bejder & R. Williams, eds, *Whale-Watching: Sustainable Tourism and Ecological Management*, pp. 323–36. Cambridge: Cambridge University Press.

Spector, S., Higham, J. E. S. & Doering, A. (2017). 'Beyond the biosphere: tourism, outer space, and sustainability'. *Tourism Recreation Research*, **42**(3): 273–83.

Statistics New Zealand (2018). *Kaikōura Recovery by the Numbers*. Available at https://www.stats.govt.nz/reports/kaikoura-recovery-by-the-numbers (accessed 9 November 2018).

Stevenson, J., Becker, J., Cradock-Henry, N., Johal, S., Johnston, D., Orchiston, C. & Seville, E. (2017). 'Economic and social reconnaissance: Kaikōura earthquake 2016'. *Bulletin of the New Zealand Society for Earthquake Engineering*, **50**(2): 343–51.

Strickland-Munro, J. K., Allison, H. & Moore, S. A. (2010). 'Using resilience concepts to investigate the impacts of protected area tourism on communities'. *Annals of Tourism Research*, **37**(2): 499–519.

Walker, B., Holling, C. S., Carpenter, S. R. & Kinzig, A. (2004). 'Resilience, adaptability and transformability in social-ecological systems'. *Ecology and Society*, **9**(2): 5.

Wall, G. (1993). *Impacts of Climate Change for Recreation and Tourism in North America*. Washington, DC: Office of Technology Assessment, US Congress.

Wall, G., Harrison, R., Kinnaird, V., McBoyle, G. & Quinlan, C. (1986). 'The implications of climate change for camping in Ontario'. *Recreation Research Review*, **13**(1): 50–60.

Wheeller, B. (1991). 'Tourism's troubled times: responsible tourism is not the answer'. *Tourism Management*, **12**(2): 91–6.

Wheeller, B. (1995). 'Egotourism, sustainable tourism and the environment – a symbiotic, symbolic or shambolic relationship?' In A. V. Seaton, ed., *Tourism: The State of the Art*, pp. 647–54. Brisbane: John Wiley and Sons.

Wilson, J., Stewart, E., Espiner, S. & Purdie, H. (2014). '"Last chance tourism" at the Franz Josef and Fox Glaciers, Westland Tai Poutini National Park: stakeholder perspectives'. *LEaP Research Report Number 34*. Lincoln University, New Zealand.

World Commission on Environment and Development (1987). *Our Common Future*. New York: Oxford University Press.

Young, M., Higham, J. E. S. & Reis, A. (2014). '"Up in the air": a conceptual critique of flying addiction'. *Annals of Tourism Research*, **49**: 51–64.

Young, M., Markham, F., Reis, A. & Higham, J. E. S. (2015). 'Flights of fantasy: A theoretical reformulation of the flyers' dilemma'. *Annals of Tourism Research*, **54**: 1–15.

4 Understanding how context affects resilience and its consequences for sustainable tourism

Johanna Loehr, Cherise Addinsall and Betty Weiler

This chapter builds on the thinking and frameworks of previous literature and this book's earlier chapters, to examine how context can both constrain and facilitate resilience and sustainability in tourism. We use Vanuatu as an example to illustrate the theoretical concepts and finalise the chapter with a research agenda for future studies in this area.

Definitions, concepts and frameworks

An underlying assumption of this chapter is that tourism often has a high reliance on the natural and cultural environment, at scales ranging from a particular attraction (e.g. a national park or coastal/marine reserve) to a region or landscape (e.g. an island or mountain range) to an entire political entity (e.g. a state, province or country). As it facilitates interaction between visitors and the places they come to visit, a tourism destination can thus be regarded as a complex socio-ecological system (SES), also described as a human-environmental system (Ostrom, 2009), in that it includes both natural and societal resources (Becken, 2013). There has been increasing interest and demand in applying systems and socio-ecological approaches in tourism research (Mai & Smith, 2015; Ruiz-Ballesteros, 2011), highlighting the need to integrate the planning and management of the socio-cultural and the natural environmental elements of tourism (Heslinga, Groote & Vanclay, 2017).

When defining tourism as an SES, it should be acknowledged that tourism is a subsystem of a larger system (Boguslaw, 2001) and that it can be split into ever smaller subsystems, also referred to as system hierarchy (Skyttner, 1996). Therefore, the human and environmental elements of any destination will influence and be influenced by links to systems at scales above and below it in the hierarchy. In this chapter, we use the term scale to refer to a range between the broader system, for example at an international level, through to a narrow (e.g. local) system. Lew (2014) defined scales of tourism interest scaling from that of the entrepreneur to those that are publicly shared. Pressures for change occur at a variety of scales; some may impact an individual entrepreneur, while others may impact an entire community or social group (Lew, 2014). The appropriate scale for analysis of a tourism system should be chosen based on the problem under investigation.

Like any complex SES, the capacity for resilience, i.e. the ability to react to unexpected or unpredictable shocks and changes, is necessary for the effective functioning and development of tourism destinations (Ruiz-Ballesteros, 2011) and the capacity for sustainability for conserving what is valued by community members (Lew et al., 2016). A key element of managing for both sustainable and resilient destinations (see Lew et al., 2016 for a discussion on the two concepts) is understanding *context*, particularly the underlying processes and social interactions that take place between stakeholders at different scales of the system, as these influence stability and the capacity to adapt to change. By context we mean any number of biophysical, geographical, political, cultural or social factors, and systems that independently and collectively distinguish a particular tourism destination. Of course, the influence of context on sustainability and resilience will be different at different scales.

Of particular relevance to this chapter's focus on context is the need to understand power and power relations, culture, and "cultural values, historical context and ethical standpoints of the kinds of actors involved" (Cote & Nightingale, 2012, p. 480). Understanding goals and underlying values helps determine preferred resilience and sustainability approaches for the destination (Lew et al., 2016). For example, belief systems and practices that devalue and disempower women in decision making will compromise capacity for long-term resilience (Cohen et al., 2016), particularly for sectors that are highly reliant on female knowledge, abilities, resources and networks (e.g. the local handicraft sector in Nepal, or the beach/resort massage sector in Bali). Cultures with strong community values and extensive family ties can draw on those support networks for assistance in times of crisis, enhancing their capacity for resilience. An example is Samoa where tourism operators are strongly influenced by the Samoan belief system and way of life emphasising family values, social networks and reciprocity, all of which can serve as support in times of disruption (Parson et al., 2017). Assessing how these contextual characteristics are institutionalised and expressed at different scales of the tourism SES will allow understanding and harnessing of the underlying system structures.

In addition, there are forces or drivers of change to the SES that influence a host destination's sustainability and resilience in the face of evolving tourism development. Drivers can be external to the chosen scope of a tourism SES, that is, influenced by a system at a higher scale (e.g. something that happens in the broader political system), or can be internal to the system under investigation itself (e.g. a change in aviation policy or services). In order to manage a tourism SES in a way that it is not only sustainable but also resilient, any drivers and their impacts need to be understood. System thinkers have identified and discussed the notion of slow and fast variables (e.g. Crépin, 2007; Walker et al., 2012). Fast variables are usually those of concern to the users of the system, while slow variables shape fast variables and determine how these react to external drivers (Walker et al., 2012). In a tourism context, Lew (2014) introduced the scale, change and resilience (SCR) model, which highlights that change occurs at different rates and affects actors at different scales within the system, creating different contextual situations. As such, forces can occur as slow-change drivers (something that may take years or decades to have obvious and widespread impact) and fast-change drivers (an event that has a sudden and immediate impact) (Lew, 2014; Lew et al., 2016). Here, we build on this model by differentiating what happens within the system under investigation (internal), and at systems at a higher level in the hierarchy (external). An example of an external slow-change driver is climate change, while an example of an internal slow-change driver is demographic change within a destination. Slow-change drivers sometimes manifest as fast-change drivers (e.g. more frequent or extreme weather and natural events due to climate change; or sudden epidemics that are possible, more likely or more widespread due to global human population movement). A terrorist attack is an example of an external fast-change driver, while a political coup is an example of an internal fast-change driver. Of course, separation into external vs internal is not straightforward as it depends on the scale of analysis. These forces may or may not impact the SES in other locations as well, but outcomes will be different due to the context of the system, and will influence how the system reacts to those changes. For example, the response to a volcanic eruption or even a change of government will in part depend on the belief and value system of actors involved in the SES, the strength of social reciprocal networks, their perceptions of risk and their views on economic opportunity (Eiser et al., 2012). A volcanic eruption, for example, may be a problem at a local scale, but an opportunity at a broader scale.

In summary, the four-cell framework of internal/external and slow/fast drivers focuses on where change is generated and at which rate. The impact of change, whether that change ends up being a facilitator or barrier to sustainable and resilient destinations depends on scale and on contextual factors such as power, culture and value systems. Thus context plays an important part in SES planning, including tourism (Cote & Nightingale, 2012). At the very least, context will moderate how the system can and will react to internal and external forces. More fundamentally, context will determine the way the system reacts to types of change (internal/external, fast/slow): will the change driver be perceived and responded to as a barrier or as an opportunity? And what will be conserved and protected (sustainability) and what will be adapted/changed into something new (resilience) (Lew et al., 2016)? This is why the same drivers can result in different responses by a tourism destination. Table 4.1 provides examples of how context may influence sustainability and resilience under different areas of change. A thorough appreciation of the local context is therefore important for strategic tourism planning and policy interventions/strategies in order to determine and build sustainable and resilient tourism destinations.

Table 4.1 provides an over-simplification of how context determines the impacts and responses to change, as a destination might be resilient to change at one scale but not another as these hierarchical SESs are adaptive and complex (Ostrom, 2009). For example, a destination at national scale may be resilient to a change in visitor demand caused by a weather event or incident (fast-change variable)

Table 4.1 Areas of change and general examples of how context influences destination sustainability and resilience

Areas/drivers of change	Examples of context acting as a facilitator	Examples of context acting as a barrier
Change in atmospheric/ biophysical condition	Warming climate making it more attractive for tourism	Tropical climate that is subject to higher risk of extreme weather events
Change in biodiversity and landscape condition	Highly diverse natural environment, fertile soil/optimal agriculture conditions may contribute to resilience	Limited natural resources may contribute to vulnerability
Change in public health condition	Easy access to health care may contribute to resilience	Limited health care infrastructure may contribute to vulnerability
Change in social condition	Strong social networks may contribute to resilience	Weak social networks may make it vulnerable to civil unrest, inequality, vulnerability
Change in economic condition, livelihoods	Access to variety of markets, trade agreements/partners may contribute to resilience	Increasing access to a variety of markets may result in breakdown of informal traditional economic activities, unsustainable use of resources, and impact social reciprocal systems which may contribute to vulnerability
Change in cultural condition, ethnicity	Strengthening of custom and traditional knowledge systems can promote sustainable use of resources, sustainable farming practices and support social reciprocal systems which may contribute to resilience	Breakdown of traditional governance systems can increase land disputes, unsustainable use of resources, inequality, and impact informal social safety nets which may contribute to vulnerability

whereas at the local level it is vulnerable. In addition, the specific context of an SES also influences how systems at different scales are linked, such as through forms of governance determined by the political context, forms of social interaction and cultural norms.

Vanuatu's context, the drivers of change and contextual factors

This chapter uses a case study of Vanuatu, a South Pacific Island state where tourism plays an important and growing role in its developing country economy. Vanuatu is located in the Southwest Pacific Ocean, 1,770km east of Australia and 800km west of Fiji (Fingleton, 2005). Vanuatu has a total land area of 12,190 square km encompassing 82 islands of volcanic origin (Cheer, 2013). The country is divided into six provinces and has a population of approximately 270,000 people, concentrated on 16 main islands. The majority of the population consists of Ni-Vanuatu (people of Vanuatu) with over 80 per cent residing in rural areas. There are 113 indigenous languages spoken in Vanuatu in addition to Bislama (the national language), English and French. Vanuatu gained independence in 1980 and has been classed as a least developed country (LDC) by the United Nations (AusAID, 2011). Remoteness, inaccessibility, isolation and few economic resources affect the capacity of Vanuatu to shift out of its position as an LDC (Prasad & Giacomelli, 2012).

The United Nations labelling of Vanuatu as an LDC is fraught with contextual issues. In response to this there has been a call for a greater focus on the role of cultural, social and environmental resources as key components to resilience (Christensen & Mertz, 2010), and more empirical and interdisciplinary studies that specifically target small island dynamics and link social-economic and ecological processes. This is particularly relevant for Vanuatu as it is among the last remaining countries in the South Pacific where residents' livelihoods rely principally on the traditional economy, i.e. a subsistence economy that is based on a traditional governance system and utilises kinships networks and customary land ownership to access food and other resources, often at lower system scales such as the community level. Regenvanu (2009) highlights the importance of the traditional economy as a source of resilience for Vanuatu's population and a buffer from uncertainty inflicted by the global economy. This contrasts and often clashes with the cash economy which encompasses a much larger system scale and in which "... the Western lifestyle and the capitalist economy are premised on ignoring the community (individualism), unsustainable use of the environment (consumerism) and removing food and social security – all basic foundations of the traditional economy" (Regenvanu, 2007, p. 1).

Addinsall (2017, p. 23) notes the importance of "encouraging sustainable economic activity at the individual and household level while operating within reciprocal networks of exchange and obligation at the community level", thereby acknowledging the importance of underlying cultural and community structures at a lower system scale. Therefore, more work is needed to understand the local context in Vanuatu, in particular the role of informal and formal institutions and how best to merge cultural and custom values, the cash and traditional economies and governance as a precursor for resilience.

Tourism in Vanuatu

Tourism has been held up as the biggest contributor to Vanuatu's gross domestic product (GDP); however, Ni-Vanuatu receive less than 10 per cent of the total tourism expenditure (Stefanova, 2008). In addition, a variety of internal and external drivers of change impact tourism in Vanuatu (Table 4.2). The government's priority on foreign investment in the tourism industry as well as the lack of control on

Table 4.2 Drivers of change impacting tourism in Vanuatu

	Fast	Slow
Internal	– Upgrading of facilities and infrastructure – Volcanic eruption – Pollution due to poor waste management practices	– Demographic change – Tourism policy and planning initiatives – Shift from subsistence to cash economy
External	– Extreme weather events such as cyclones – Cruise industry decisions and actions – Foreign investment in targeted initiatives	– Colonialism and Westernisation – Climate change – Increased destination competitiveness – International standards and rising visitor expectations

capital flows and exchange rates (due to an open economy) has resulted in non-inclusive development, increased economic inequalities, dispossession of land and impacts on social systems (Stefanova, 2008), the latter being crucial to resilience in the absence of formal safety nets.

Tourism in Vanuatu relies on images of turquoise water, white sandy beaches, palm trees, a strong culture and smiling friendly faces (Harrison, 2004), with Ni-Vanuatu being portrayed as the 'happiest people on earth' (BBC, 2006; Happy Planet Index, 2016). However, these images mask the real economic, environmental, social and governance issues in Vanuatu. External and internal drivers risk negatively impacting on the very qualities that attract people to visit Vanuatu. For example, the tourism industry in Vanuatu is concentrated in the capital city of Port Vila; exhibits high leakage, foreign dominance and economic inequalities; relies heavily on imported foods; has resulted in dispossession of land; is responsible for large-scale vegetation clearance, dredging of mangroves and estuaries, and coastal pollution; and is largely unregulated (Stefanova, 2008). These factors demonstrate that the government's commitment to developing sustainable tourism in Vanuatu is often not realised (Harrison, 2004). However, "there are difficulties in estimating the extent of these issues due to the lack of reliable data on tourism impacts in countries [such as Vanuatu]" (Addinsall, 2017, p. 50).

Some of these internal and external drivers are particularly critical in driving the need for increased government regulation of the tourism industry in Vanuatu. Acknowledging context-specific characteristics such as alternative world views and knowledge systems is vital when considering sustainable tourism development in Vanuatu. Studies have found a positive relationship between resilience and a feeling of well-being among communities that are operating in hybrid economies that recognise both formal and informal processes (VNSO, 2012). With the absence of formal safety nets in Vanuatu, informal community and culturally based social protection systems occurring at a lower system scale are vital to ensure resilience and protect particularly vulnerable people and environments from external change drivers such as macro-economic shocks and the impacts of climate change.

Strategies focused on growth notoriously exclude environmental and social benefits and costs in their analyses, whereas traditional economic activities are often based on sustainability (Anderson, 2011). Allen (2008) suggests the continued push for Vanuatu to develop a capitalist economy, with the introduction of a system of freehold or individual title, and deregulation of the market, could lead to sharp socio-economic differentiation and, in the long term, the emergence of a landless peasantry. The approach to tourism in Vanuatu has largely supported a capitalist approach. Therefore, some government intervention and industry regulation, together with a genuine recognition of local context, such as informal and traditional internal processes, are needed to prevent the tourism industry from passing on the costs of environmental degradation, cultural commodification and social displacement to the most vulnerable (Addinsall et al., 2016).

Tourism policy making and planning responses to foster sustainability and resilience

The Vanuatu government (like many South Pacific governments) is now reconsidering how tourism has been approached in the past (operating within a free market ideology) and assessing the need for government intervention and industry regulation (Schilcher, 2007). However, it can be argued that for tourism to truly contribute to sustainable development and be therefore labelled as 'sustainable' (within a Vanuatu context) it needs to better incorporate local context and align with Ni-Vanuatu values, focusing on equity instead of growth. This would require an ideological change to how tourism is developed, moving from a focus on growth to alternative economic ideologies based on Ni-Vanuatu world views, such as community benefits from tourism investments.

With the development of national policies and plans based on sustainability, Vanuatu's approach in recent years is often labelled as 'sustainable'. For example, the release of the National Sustainable Development Plan (NSDP) (2016–30) as the highest level policy framework is an attempt to further extend linkages between resources, policy and planning to Vanuatu's culture, custom, traditional knowledge and Christian principles putting forward cultural heritage as the foundation of an inclusive society. The NSDP demonstrates a strong attempt to address the disconnection between policy and custom, using phrases such as "Cultural heritage as the foundation of an inclusive society", "traditional and social safety nets", and "balance the interface between formal and traditional governance systems". However, separating the NSDP goals into 'society', 'environment' and 'economy' pillars fails to apply a holistic view on system boundaries. A review of governmental policies and documents alongside communications with government and tourism industry stakeholders reveals an increasing focus on growing the tourism industry and export sectors in Vanuatu with concern expressed as to the ability of national environmental, tourism planning and investment legislation to ensure growth is sustainable and equitable. There is also limited evidence of the NSDP influencing governmental decisions surrounding tourism development (Spooner, 2018). For example, within the tourism-related indicators of the NSDP there is no mention of tourism's role in protecting environmental resources; providing opportunities for vulnerable groups of people; increasing the level of local ownership of tourism businesses; protecting and respecting culture; and sustainable waste, water and energy management.

Ensuring a sustainable, inclusive and resilient tourism industry requires a refocus on environmental and social factors (Addinsall, Weiler & Spooner, 2018), an approach which is gaining traction in Vanuatu (Addinsall et al., 2016). Both the tourism industry and the country are highly susceptible to internal and external drivers of change (Table 4.2). To gain a better understanding of how these factors interact and at different system scales, local context needs to be more thoroughly considered.

The government in Vanuatu has been pressured for some time by external forces emphasising neoliberal discourses of 'growth', 'efficiency', 'reform', and 'governance', while only a handful of non-governmental organisations and researchers are challenging current economic and trade policy arrangements in the South Pacific (Addinsall, 2017; Anderson & Lee, 2010; Regenvanu, 2009; Simo, 2010). The current system, as well as change drivers, make it virtually impossible for Ni-Vanuatu to compete with foreign investors in the tourism industry, with evidence of few benefits from the tourism industry flowing to Ni-Vanuatu.

Actions taken to deal with change drivers through understanding context

For some time the Vanuatu Department of Tourism (DoT) has acknowledged the need to develop policies and plans to regulate the tourism industry. A key driver for increased regulation in Vanuatu is the government's desire to see increased local participation in the tourism industry, particularly in the outer islands. However, with this increase in local participation, issues have arisen due to the manifestation of negative reviews by visitors following their holidays, with respect to the lack of infrastructure, poor customer service and substandard hygiene.

Implementing regulation of the tourism industry requires a determination of what the needs of Ni-Vanuatu and stakeholders are at a smaller scale and how to align them with meeting sustainable social, economic, cultural and environmental goals. At a larger scale, the country as a system can be resilient and sustainable, while the sustainability of subsystems at smaller scales, such as in the outer islands, or at the community level is compromised. The development of the Vanuatu Sustainable Tourism Policy (VSTP) (2019–30) (Department of Tourism, 2018) and mandatory tourism sector standards are seen as ways of responding to these and other external and internal forces. These regulation processes seek to meet key environmental, social, cultural and economic objectives at a range of scales: the international level (sustainable development goals (SDGs); regional level (South Pacific Tourism Organisation (SPTO)); national level (NSDP); and sectorial level (Vanuatu Strategic Tourism Action Plan (VSTAP) 2014–18) (Ministry of Tourism, Industry, Commerce & Ni-Vanuatu Business, 2013).

To be specific, the Vanuatu government has seen a need for policy guidelines, the implementation of baseline standards in tourism products, enhanced service quality and the management of risk, all aimed at achieving long-term sustainability. As a result, the government passed the Tourism Councils Act (2012) to allow for the introduction of the VSTP and mandatory tourism sector standards.

The VSTP provides a guiding framework and direction for the government of Vanuatu and all stakeholders to develop their tourism sector in a sustainable manner. The VSTP is a "living" and flexible document that represents a stepping stone of what will be an ongoing consultative process of incorporating key messages from stakeholders across Vanuatu (Addinsall, Weiler & Spooner, 2018), aligning and realigning policy across government sectors where required, and responding to slow and fast, internal and external drivers. Collaboration between tourism and other sectors as well as adaptability between policies and plans are both essential in order to support sustainability initiatives and linkages that strengthen the tourism sector and Vanuatu as a country. The VSTP focuses on enhancing the resilience of Vanuatu's SES including its cultural, social and ecological systems in the face of the changes, complexity and uncertainty that typify twenty-first-century Vanuatu.

It needs to be noted that many previous attempts had been made to develop policy and standards in Vanuatu, with minimal success in implementation (Spooner, 2018). In response to these failed attempts to introduce policy and international standards programmes, the Vanuatu government stressed the need for a policy and programme that would fit the Vanuatu context and that stakeholders at all levels would embrace. DoT also changed its approach of employing largely external, foreign aid-funded, short-term consultants to develop the policy and standards, to funding a highly skilled local (Ni-Vanuatu) in the role of full-time Principal Accreditation Officer, who would be charged with leading both the policy development process and the development of tourism standards. Another key difference is that the approach taken by the Principal Accreditation Officer (Jerry Spooner) in developing the policy and standards was to embrace a very high level of stakeholder engagement. A wide range of stakeholders from the tourism industry, central and local government, community, and local cultural groups in Vanuatu were engaged in the development of the policy and standards over a two-year time frame. Consultation with outer island stakeholders is continuing. This example highlights how the complex adaptive SES that is Vanuatu tourism is organising itself, after a few failed attempts, to maintain its key function and identity, a sign of system resilience (Walker et al., 2012).

The Tourism Standards were developed under a programme entitled The Vanuatu Tourism Permit and Accreditation Program (VTPAP), launched in 2015 by the

DoT to ensure tourism operators are operating at, or above, minimum standards throughout Vanuatu's tourism sector. A focus of the programme is that Vanuatu is perceived by visitors as a safe and reliable destination that showcases Ni-Vanuatu culture and identity as well as the country's natural assets as tourism attractions. The VTPAP seeks to ensure that quality service and facilities are being provided to all tourists and to build safety and reliability into Vanuatu's tourism brand (Spooner, 2015).

The VTPAP can be seen as a first step in supporting the delivery of higher quality tourism experiences, and allows Vanuatu to be more competitive in the South Pacific leading to enhancement of the Vanuatu economy through translation into increased visitor numbers. VTPAP is customised to all tourism businesses in both urban and rural areas/outer islands. It provides incentives through promotional, marketing, training and capacity-building opportunities for these tourism businesses. It also encourages dispersal of tourists to these rural areas which creates spillover effects, such as expenditure in other local businesses. Increasing numbers of tourist arrivals directly links to the VSTP and NSDP objectives. In terms of GDP, VTPAP's purpose is to protect the industry and increase revenue—it supports tourism stakeholders that have had little engagement in the formal economy managing other businesses, in a way that is competitive while still supporting their traditional roles.

A contextual approach: power, culture and values in the Vanuatu context

Consideration of the local context in the consultation process of the VSTP and the VTPAP development and implementation enabled voices of stakeholders that are commonly missing in the development of government policy, plans and actions, to be heard. Examples of this were holding meetings in Bislama, managing outspoken dominant people in workshops, and organising one-on-one interviews with stakeholders who were not comfortable contributing in workshop environments. This wide consultation allowed the government to ensure that consideration of local cultural nuances (previously seen as inhibitors to government plans and strategies) were included when the VSTP and VTPAP were operationalised.

Early in the process of developing the VTPAP it became evident that simply taking an existing standard or accreditation programme and applying it to the tourism sector in Vanuatu was not possible: the cultural and economic context required that the international tourism standards be adapted. Several strategies and approaches ensued. Some key lessons emerge from this case study that relate to context. Firstly, the large scale buy-in to the programme to date can be attributed to the investment of time in the development and implementation of the VTPAP. Secondly, employing a Ni-Vanuatu (i.e. not a foreign national) as Principal Accreditation Officer, with language, cultural and social skills, knowledge and approaches that are appropriate to the context, proved critical. As an educated multi-lingual with a deep understanding of the complexities of tourism operations in Vanuatu, Spooner was able to gain the trust and support of both locals and expatriate tourism operators, government personnel, appropriate authorities within communities and stakeholders of all ages and gender. Thirdly, extensive engagement with stakeholders following proper custom protocol helped ensure a better acknowledgement of context and thus the creation of a destination-specific programme that is relevant and workable in the Vanuatu tourism industry. Finally, this stakeholder engagement was largely facilitated by embracing and engaging with existing power-brokers, namely, functioning and active tourism associations and local government tourism councils in each province of Vanuatu. Making tourism permits conditional on being a member of an active tourism association further streamlined stakeholder engagement activities.

Through the Decentralisation and Local Government Regions Act No.1 of 1994, more power was allocated to provincial governments to manage their provinces autonomously; however, the structure of Vanuatu's national government still heavily influences decision making at provincial level. As a result of colonialism, which has been an external slow driver of change, Vanuatu's national government operates within a Western-influenced manner even though Vanuatu gained its independence in 1980. This presents challenges when internal drivers, such as the informal, traditional economy and customs, intersect with government.

In response to these challenges, the VTPAP sought to provide more influence to both provincial government systems and Vanuatu's custom laws aligning more to the Decentralisation Act. This was achieved by working within provincial, custom and community structures to develop the standards to meet their local context, rather than a top-down approach to development and implementation. The development of the VTPAP operational plan has highlighted that every province manages their standards programme suited to their context, ensuring that custom governance, a provincial local government structure, and the informal traditional economy is taken into consideration. For example, in Tafea province the majority of tourism products are located outside municipal areas (central business areas) and so fall under the responsibility of the provincial government. This is in stark contrast to Shefa province where the majority of tourism businesses are located within the capital city. Thus, in Tafea, tourism permits are signed off by the Secretary General before approval by the national government (Director of Tourism), whereas in Shefa the national government directly signs off on all tourism permits. Custom and cultural considerations were also considered in the VTPAP by incorporating criteria such as written consent from area chiefs for any tourism permit renewals and for new permit applications.

The ongoing continuous improvement plan for VTPAP together with the roll out of the VSTP are internal slow drivers that respond to external drivers such as the development of the Global Sustainable Tourism Council (GSTC) criteria for hotels and tours as well as destinations. Meeting global standards makes it possible and appropriate for Vanuatu to work towards becoming a GSTC-certified destination. This continuous improvement will seek to attract quality visitors—higher yield, longer stay, culturally aware and environmentally responsible, while at the same time providing tourism operators with guidelines to reduce their organisation's impact on the natural and socio-cultural environment. As such, the VSTP and the VTPAP frame sustainable tourism not as a discrete set of tourism products, but as something for the entire tourism and non-tourism sectors to work towards. The premise of the VSTP and the VTPAP is that Vanuatu will thrive and survive when all forms of tourism strive to be more responsible and sustainable.

Research agenda

The previous sections have highlighted by way of a case study of Vanuatu how policy making and planning can better harness context to build resilience and sustainability. In addition, this case study helps uncover where further research is needed both in the South Pacific context and elsewhere.

Vanuatu, the South Pacific and other contexts

More work is needed to understand the role of informal and formal institutions and how best to merge cultural and custom values, the traditional and cash economies, and governance as a precursor for resilience.

The Vanuatu Tourism Permit and Accreditation Program is only a first step. Further research is needed to better understand how long-term sustainability and resilience can be achieved and implemented at all social and geographical scales.

Due to the specific nature of the tourism sector in Vanuatu and other South Pacific islands where tourism, communities and the environment are strongly interlinked, more empirical and interdisciplinary studies are needed that specifically target such small island contexts and link social-economic and ecological processes to identify context-specific solutions to questions related to resilience and sustainability.

Tourism destinations, change and resilience

We have identified different types of change drivers (e.g. internal vs external; slow vs fast change) impacting Vanuatu's tourism sector. To ensure resilient and sustainable tourism destinations can be built and maintained, consideration of the following questions will lead to a better understanding of what is needed for tourism SES:

- What variables make up the system, what are the dominant processes that influence decision making and what are the context-specific characteristics that influence system dynamics?
- How do these change types influence tourism and tourism stakeholders at different scales of the system and what does that mean for the overall resilience of the system?
- How do and how should responses differ between reacting to fast and slow

change drivers (e.g. disaster risk reduction vs building long-term adaptive capacity)?

- When we think about tourism resilience, to what degree do we need to consider lower scale systems and the larger scale system?
- How can tourism be a tool to enhance resilience of the wider system?

References

Addinsall, C. 2017. 'Agroecology and sustainable rural livelihoods: interdisciplinary research and development in the South Pacific'. PhD thesis, Southern Cross University, Lismore.

Addinsall, C., Scherrer, P., Weiler, B. & Glencross, K. 2016. 'An ecologically and socially inclusive model of agritourism to support smallholder livelihoods in the South Pacific'. *Asia Pacific Journal of Tourism Research,* **22**(3): 301–15.

Addinsall, C., Weiler, B. & Spooner, J. 2018. 'Vanuatu Sustainable Tourism Policy, Appendices'. Vanuatu Department of Tourism.

Allen, M. 2008. 'Land reform in Melanesia: state, society and governance in Melanesia, Briefing Note, Number 6'. Canberra: ANU Press.

Anderson, T. 2011. 'Melanesian land: the impact of markets and modernisation'. *Journal of Australian Political Economy,* **68**: 86–107.

Anderson, T. & Lee, G. 2010. 'Introduction: understanding Melanesian customary land'. In T. Anderson & G. Lee, eds, *In Defence of Melanesian Customary Land,* pp. 2–4. Sydney: AID Watch.

AusAID. 2011. 'Australia and the least developed countries: partners in development'. Available at https://dfat.gov.au/about-us/publications/corporate/annual-reports/ausaid-annual-report-2010-2011/02performance.html (accessed 30 December 2013).

BBC. 2006. 'What's so great about living in Vanuatu?' Available at http://news.bbc.co.uk/2/hi/uk_news/magazine/5172254.stm.

Becken, S. 2013. 'Developing a framework for assessing resilience of tourism sub-systems to climatic factors'. *Annals of Tourism Research,* **43**: 506–28.

Boguslaw, R. 2001. *Systems Theory Encyclopedia of Sociology,* Vol. 5. 2nd edn., pp. 3102–6. New York: Macmillan Reference USA.

Cheer, J. M. 2013. 'Outer island tourism in the South Pacific and the Millennium development goals: understanding tourism's impacts'. In K. S. Bricker, R. Black & S. Cottrell, eds, *Sustainable Tourism and the Millennium Development Goals: Effecting Positive Change,* pp. 23–36. Burlington, VT: Jones and Bartlett Learning.

Christensen, A. E. & Mertz, O. 2010. 'Researching Pacific Island livelihoods: mobility, natural resource management and nissology'. *Asia Pacific Viewpoint,* **51**(3): 278–87.

Cohen, P. J., Lawless, S., Dyer, M., Morgan, M., Saeni, E., Teioli, H. & Kanor, P. 2016. 'Understanding adaptive capacity and capacity to innovate in social-ecological systems: applying a gender lens'. *Ambio,* **45**(3): 309–21.

Cote, M. & Nightingale, A. J. 2012. 'Resilience thinking meets social theory'. *Progress in Human Geography,* **36**(4): 475–89.

Crépin, A.-S. 2007. 'Using fast and slow processes to manage resources with thresholds'. *Environmental and Resource Economics,* **36**(2): 191–213.

Department of Tourism. 2018. *Vanuatu Sustainable Tourism Policy 2019–30.* Ministry of Tourism, Industry, Commerce & Ni-Vanuatu Business, Vanuatu: Port Vila.

Eiser, J. R., Bostrom, A., Burton, I., Johnston, D. M., McClure, J., Parton, D., van der Pligt, J. & White, M. P. 2012. 'Risk interpretation and action: a conceptual framework for responses to natural hazards'. *International Journal of Disaster Risk Reduction,* **1**: 5–16.

Fingleton, J., 2005. 'Introduction, privatising land in the Pacific: a defence of customary tenures'. Discussion Paper No. 80. Canberra: The Australia Institute.

Happy Planet Index. 2016. 'Vanuatu'. Available at http://happyplanetindex.org/countries/vanuatu.

Harrison, D. 2004. 'Tourism in Pacific Islands'. *Journal of Pacific Studies*, **26**(1–2): 1–28.

Heslinga, J. H., Groote, P. & Vanclay, F. 2017. 'Using a social-ecological systems perspective to understand tourism and landscape interactions in coastal areas'. *Journal of Tourism Futures*, **3**(1): 23–38.

Lew, A. A. 2014. 'Scale, change and resilience in community tourism planning'. *Tourism Geographies*, **16**(1): 14–22.

Lew, A. A., Ng, P. T., Ni, C.-C. & Wu, T.-C. 2016. 'Community sustainability and resilience: similarities, differences and indicators'. *Tourism Geographies*, **18**(1): 18–27.

Mai, T. & Smith, C. 2015. 'Addressing the threats to tourism sustainability using systems thinking: a case study of Cat Ba Island, Vietnam'. *Journal of Sustainable Tourism*, **23**(10): 1504–28.

Ministry of Tourism, Industry, Commerce & Ni-Vanuatu Business. 2013. *Vanuatu Strategic Tourism Action Plan (VSTAP) 2014–18*. Ministry of Tourism, Industry, Commerce & Ni-Vanuatu Business, Vanuatu: Port Vila.

NSDP. (2016–30). 'National sustainable development plan, Vanuatu 2030, The Peoples Plan'. Republic of Vanuatu.

Ostrom, E. 2009. 'A general framework for analyzing sustainability of social-ecological systems'. *Science*, **325**(5939): 419–22.

Parson, M., Brown, C., Nalau, J. & Fisher, K. 2017. 'Assessing adaptive capacity and adaptation: insights from Samoan tourism operators'. *Climate and Development*, **10**(7): 644–63.

Prasad, B. C. & Giacomelli, A. 2012. *Trade Policy Framework, Department of External Trade (DET) & Ministry of Trade, Commerce, Industry, and Tourism (MTCIT)*. Vanuatu: Port Vila.

Regenvanu, R. 2007. 'The traditional economy – what's it all about?' *MILDA*. Available at http://milda.aidwatch.org.au/resources/documents/traditional-economy-whats-it-all-about.

Regenvanu, R. 2009. 'The traditional economy as a source of resilience in Vanuatu'. In T. Anderson & G. Lee, eds, *In Defence of Melanesian Customary Land*, pp. 30–34. Sydney: AID/WATCH.

Ruiz-Ballesteros, E. 2011. 'Social-ecological resilience and community-based tourism'. *Tourism Management*, **32**(3): 655–66.

Schilcher, D. 2007. 'Growth versus equity: the continuum of pro-poor tourism and neoliberal governance'. *Current Issues in Tourism*, **10**(2–3): 166–93.

Simo, J. 2010. 'Land and the traditional economy: "Your money, my life". Hu i kakae long basket blong laef?' In T. Anderson & G. Lee, G. (eds), *In Defence of Melanesian Customary Land*, pp. 40–44. Sydney: AID/WATCH.

Skyttner, L. 1996. 'General systems theory: origin and hallmarks'. *Kybernetes*, **25**(6): 16–22.

Spooner, J. 2015. *Information Pack, Vanuatu Tourism Permit and Accreditation Program*. Vanuatu Department of Tourism.

Spooner, J. 2018. 'Interview', Vanuatu Department of Tourism, Port Vila, 6 June 2018.

Stefanova, M. 2008. 'The price of tourism: land alienation in Vanuatu'. *Justice for the Poor – Briefing Notes January 2008*, **2**(1). Available at http://documents.worldbank.org/curated/en/862021468245413815/pdf/436860BRI0J4P010B0x327368B01PUBLIC1.pdf (accessed 20 January 2012).

VNSO. 2012. 'Alternative indicators of well-being for Melanesia, Vanuatu pilot study report'. Vanuatu National Statistics Office (VNSO), Malvatumauri National Council of Chiefs, Port Vila, Vanuatu. Available at http://www.vnso.gov.vu/ (accessed 12 October 2015).

Walker, B. H., Carpenter, S. R., Rockstrom, J., Crépin, A.-S. & Peterson, G. D. 2012. 'Drivers, "slow" variables, "fast" variables, shocks, and resilience'. *Ecology and Society*, **17**(3): 30.

5 Concepts for understanding the visitor experience in sustainable tourism

Zachary D. Miller, William L. Rice, B. Derrick Taff and Peter Newman

Introduction

The visitor experience is a user-constructed concept that is facilitated by a dynamic interaction between the social, ecological and managerial components that exist in parks and protected areas (Borrie & Roggenbuck, 1998; Manning, 2011). Social elements include people, their psychological states, their behaviors, and the activities they engage in (Borrie & Roggenbuck, 1998). Ecological elements include the natural components of a protected area, like wildlife, water features, scenic vistas, natural soundscapes, degree of modification of the landscape, and night skies. Managerial elements include regulations and restrictions, transportation systems, trail systems, and other facilities and services. The synergistic product of the social, ecological and managerial components of a protected area is the visitor experience (Manning, 2011).

One of the core responsibilities of the sustainable tourism industry is to understand and manage for visitor experiences. The challenge for sustainable tourism in relation to visitor experiences is not only the identification of desired experiences and outcomes to promote tourism opportunities, but also the evaluation of the effectiveness of management actions related to managing visitor use. This chapter provides guidance for understanding and managing visitor experiences in the context of sustainable tourism. First, we discuss key concepts for understanding visitor experiences: (a) recreation experience preferences and (b) outcomes-based management. These two sections examine why tourists visit parks and protected areas, and the types of cultural ecosystem services, or benefits, they obtain from their experiences. After laying this foundation, we develop a five-step Management-By-Objectives framework as a way to manage for quality visitor experiences. Next, we identify gaps in the research and how inquiries into these areas can contribute to sustainable tourism development. Lastly, we describe a path forward for continuing to provide high quality visitor experiences in sustainable tourism.

Why are visitor experiences important to sustainable tourism?

Before further discussing visitor experiences and their management, it should first be recognized that providing quality visitor experiences is critical to sustainable tourism. This is because quality visitor experiences generate a variety of additional benefits for parks and protected areas (McCool, 2006). For instance, quality visitor experiences lead to an increase in return visitors and recommendations to others to visit the area (Castellanos-Verdugo et al., 2016). In turn, the sustained visitor use from return visitors and recommendations adds to the economic sustainability of the area (Moore, Rodger & Taplin, 2015; Pinkus et al., 2016). This includes contributions like jobs (Ardoin et al., 2015), reducing poverty (Snyman, 2017), and supporting rural economies (Valdivieso, Eagles & Gil, 2015). In addition to community-centered economic contributions, quality visitor experiences can impact conservation-related behaviors of visitors. For instance, positive and impactful memories about an experience increases park support among visitors (Jorgenson & Nickerson, 2015; Jorgenson et al., 2018). Additionally, quality visitor experiences can increase philanthropic donations to conservation organizations (Ham, 2013; Powell & Ham, 2008) and the pro-environmental behaviors of visitors (Han, Lee & Hwang, 2016). In essence, providing quality visitor experiences is necessary for the long-term sustainability of parks and protected areas (Romagosa, Eagles & Duitschaever, 2012).

Visitor experience preferences and outcomes

Visitors often have preferences for certain types of experiences, such as solitude or togetherness. This is because people believe that by engaging in an experience, they can achieve a specific goal, or outcome (Driver & Brown, 1975; Manfredo, Driver & Tarrant, 1996). The connection between visitor preferences and related outcomes is a critical concept for understanding the visitor experience. In the section below, we discuss recreation experience preferences and outcomes-focused management as approaches to understanding the visitor experience.

Recreation experience preferences

People come to parks and protected areas with specific motivations for engaging in activities and experiences. These motivations are essential for understanding the visitor experience as they provide insights into visitor behaviors and help managers provide opportunities that satisfy visitor desires (Manfredo, Driver & Tarrant, 1996). One way to think about visitor motivations is through Recreation Experience Preferences (REPs) (Driver 1976; 1983). Instead of understanding visitor experiences through activities, such as hiking, birdwatching or photography, REPs focus on the psychophysiological dimensions of the visitor experience (Manfredo, Driver & Tarrant, 1996). Rooted in motivation theory, REPs include concepts such as risk-taking, learning, experiencing nature, physical exercise and family togetherness (Driver, 1983; Manfredo, Driver & Tarrant, 1996). These dimensions of the experience help explain why visitors are coming to a destination and why they are engaging in certain activities. REPs also help explain variation among visitors within the *same* activity. For instance, although many visitors may participate in hiking, some may choose experiences that provide solitude, while others may choose experiences that provide an opportunity for family togetherness. These within-group differences further highlight the need for managers of sustainable tourism to focus on the REPs related to the visitor experience instead of just the activities.

In application, REPs can be used to inform management objectives related to visitor experiences in sustainable tourism settings. For instance, if visitors to an area evaluate opportunities for solitude and pristine natural environments as major motivations for visiting, these could be incorporated into the management objectives of an area (see "Managing for Quality Visitor Experiences" below). However, these visitor preferences are only one consideration when forming management objectives, and other aspects such as ecological impacts, foundation documents and providing a diversity of experiential opportunities across a landscape also need to be considered when forming management objectives for an area.

Outcomes-focused management

REPs represent a well-tested method of measuring recreationists' expectations. Yet, to manage solely for an expectation is to fall short of the fulfillment of the expectation—the outcome. Beginning in the 1970s, social scientists began to rethink the means by which we manage protected areas towards an approach that directly manages for the realization of visitor goals (Driver & Tocher, 1970). This is commonly referred to as the Benefits Approach to Leisure (BAL). BAL follows REPs in that a visitor enters a protected area with a specific goal in mind. Using the best available guidance to inform their behavior, they select an activity and setting that will most likely lead to a certain benefit, or "bundle" of benefits.

Outcomes-Focused Management (OFM) is the most recent iteration of the BAL (Driver, 2008). This framework recognizes that experiences may result in both benefits (positive) and costs (negative). However, instead of focusing on the motivations visitors arrive with (like REPs), OFM focuses on the *outcomes* of visitor experiences. It is helpful to think of their relationship through an analogy: motivations are to preference scales as benefits and costs are to outcome domains. OFM domains range from psychological outcomes like improved mental health, to personal growth benefits like improved self-confidence. However, outcomes also include concepts that accrue to households, communities and the environment. A key difference between outcomes and motivations is the inclusion of both positive and negative outcomes. Though subtle, this represents a major breakthrough. Instead of managing only to maximize the benefits of visitors' experience, the OFM framework recognizes the capacity of managers to reduce costs. By reducing costs, managers can create opportunities that maximize the benefits related to visitor experiences.

Carrying this out in sustainable tourism, however, requires careful analysis of visitor experiences and potentially extensive implementation efforts by all levels in the tourism value chain. Managers must survey visitors to understand their motivations and current outcomes, and any gaps that exist between motivation and outcome. Taking these data, managers must implement policies and make any necessary changes to the built and human capital of their site. They must also set up a system of monitoring recreational outcomes following their modifications. Managers must evaluate how successful their changes and policies have been in the delivery of desired benefits and continue to refine their tourism experience. Finally, tour operators, restaurateurs, hoteliers and other hospitality sectors must adjust and coordinate with land managers to ensure their efforts are congruent throughout the tourism value chain.

Recreation opportunity spectrum

Zoning protected areas for potential benefits that may be accrued by a recreationist is not a new concept. As noted in Driver et al. (1987), Arthur Carhart and Aldo Leopold suggested guidelines for managing different recreational opportunities on national forest landscapes in the 1920s, and a variety of well-known land use planners, managers, and protected areas-focused researchers have suggested the need for zoned opportunities. Stankey (1977) used the term "recreation spectrum" to describe this concept, and Driver et al. (1987) discussed the evolution, assumptions and tenets of what is now known as the Recreation Opportunity Spectrum (ROS).

Before ROS, zoning opportunities largely focused on activity type. Yet Driver et al. (1987) argue that "recreation is not participation in an activity but instead is the psychological state of refreshment—of being made new or anew under conditions relating to intrinsic reward, voluntary choice, and lack of time structuring" (p. 204). Therefore, ROS as it is applied today also considers setting opportunities and experience opportunities as described in the previous section (Driver et al., 1987).

The assumptions and tenets behind ROS have been tested in a variety of circumstances since its inception (e.g., Martin, Marsolais & Rolloff, 2009). However, the challenge for managers implementing ROS is in identifying which indicators matter most and determining how to monitor and adaptively manage those indicators. This is particularly challenging as recreational patterns shift and influence behaviors, settings and associated experiences. Despite these challenges, ROS has been adopted globally in a variety of contexts. For example, Haas and Aukerman (2004) have described the Water Recreation Opportunity Spectrum for zoning recreational opportunities on waterways, and Boyd and Butler (1996) introduced the Ecotourism Opportunity Spectrum to apply these concepts beyond parks and protected areas. In Panama, the Rango de Oportunidades para Visitantes en Áreas Protegidas (ROVAP) is used to zone national parks for public use (Ministerio de Ambiente, 2015). As technology enhances the scope of our capabilities, ROS will likely continue to be useful in concept and application. For example, researchers in New Zealand have posed models that allow for automated development of ROS in a manner that reduces subjectivity and improves repeatability (Joyce & Sutton, 2009), and city and landscape planners in Norway have found ROS to be applicable to blended urban and natural settings (Gundersen, Tangeland & Kaltenborn, 2015).

Managing for quality visitor experiences

Parks and conservation areas have long been recognized for their tourism value. For instance, in 1864 when the president of the United States (then Abraham Lincoln) transferred the federally administered Yosemite Valley and Big Trees grove to the state of California, it was under "the express conditions that the premises shall be held for public use, resort, and recreation" (Yosemite Grant Act of 1864). This idea was further progressed by the national park concept when the federal government of the United States declared the Yellowstone area in the territory of Wyoming "as a public park or pleasuring-ground for the benefit and enjoyment of the people" (Yellowstone Park Act of 1872). As the concept of natural areas as destinations for tourism and public enjoyment solidified, a dual-mandate for protecting valuable resources while providing for public enjoyment emerged (Organic Act of 1916). Although this dual-mandate is often presented as competing opposites, in actuality it is a reinforcing loop; the public use and enjoyment of these places depends on the protection of the resources. In other words, the type and amount of public use and enjoyment in these areas is the same type and amount that leaves the resources unimpaired. This balance of public use and resource protection is essential to park and conservation area governance to this day.

As early as 1936, people recognized that increased or inappropriate human use can degrade the essential qualities that park and conservation areas are designed to protect (Sumner, 1936). As concerns grew about degraded resource conditions and crowded natural areas from the post-World War II tourism boom of the 1950s, the idea of *actively managing* for visitor use appeared (Wagar, 1964). From this, an entire field of study arose to answer the question: *How much and what types of visitor use can be sustained in an area before the very experiences people are seeking are unacceptably degraded?* (Sumner, 1936; Wagar, 1964). The field of study that addresses this and similar questions is known as Visitor Use Management (VUM).

Visitor use management frameworks

Visitor use management is a process for managing human use in a way that maintains or achieves specified experiences or conditions (IVUMF, 2016). Carrying capacity was an early concept that emerged to address visitor use in park and conservation areas (Wagar, 1964). However, carrying capacity was challenged as a concept for managing visitor use early on and continues to be a point of controversy in the field of VUM (Whittaker et al., 2011). The controversy with carrying capacity is that it assumes linearity; for every one person added to the system, it results in a stable and proportionally equal effect (McCool et al., 2014; McCool & Lime, 2001). As an example, as one person enters a trail, congestion is increased by a consistent amount. This is demonstrably false. For instance, at Muir Woods National Monument, visitor experience was improved through signage, not in managing for a specific number of people (Box 5.1). We can see from this case that it is not only the number of people in a system that affects visitor experiences, but also the way visitors behave. As opposed to carrying capacity that focuses on the *inputs* to a

Box 5.1: In Muir Woods National Monument (California, USA), researchers applied the VERP framework to improve visitor experiences related to soundscapes. The research revealed that human-caused sound was above the science-informed threshold for acceptable levels. Educational signage was selected as a management intervention reducing human-caused sound to a level below the acceptable threshold. After implementing the signage, monitoring revealed that the signs were effective at reducing human-caused sound and improving the visitor experience in the area. This case study highlights the value of an outputs-focused MBO to manage the visitor experience (Pilcher, Newman & Manning, 2009; Stack et al., 2011).

Photo credit Zach Miller

system, a shared component of VUM frameworks is that they focus on the *outputs* of a system. These outputs are the desired experiences or conditions visitors seek in sustainable tourism settings. In this way, VUM frameworks accommodate the complex and dynamic process that is the visitor experience.

Several different VUM frameworks were developed to address visitor experience in sustainable tourism settings (Table 5.1). These include the Limits of Acceptable Change (LAC) (McCool, Clark & Stankey, 2007; Stankey, McCool, & Stokes, 1984; Stankey et al., 1985), Visitor Impact Management (VIM) (Graefe, Kuss & Vaske 1990), Visitor Experience and Resource Protection (VERP) (Hof & Lime, 1997), and the more recent Interagency Visitor Use Management Framework (IVUMF) (IVUMF, 2016). Although there are some minor differences among the frameworks (Table 5.1), most were simply developed independently to meet the needs of a particular land management agency. More recently, the IVUMF (2016) was developed to standardize the terminology and process used among US public land agencies. However, all VUM frameworks have many more similarities than differences, and any could be used to manage visitor experiences in sustainable tourism settings.

Implementing a management-by-objectives framework

All the VUM frameworks described above are considered Management-By-Objectives (MBO) frameworks (Manning, 2011). MBO frameworks share a common goal of creating an institutionalized system for learning that provides a basis for science-based decision making for managers of sustainable tourism. Viewed collectively, MBO frameworks share five key steps that can be applied

Table 5.1 A comparison of Visitor Use Management frameworks

Visitor Use Management framework[1]	Primary applications	Unique characteristics
Limits of Acceptable Change (LAC)	US Forest Service; Wilderness	Explicitly incorporates the Recreation Opportunity Spectrum (ROS)
Visitor Impact Management (VIM)	Limited US national parks, not widely used	Smaller geographic scale (focus on sites); explicit step on identifying probable causes of impacts
Visitor Experience and Resource Protection (VERP)	US National Park Service	Park purpose incorporated into establishing management objective
Interagency Visitor Use Management Framework (IVUMF)	US public land agencies	Sliding scale to assess level of analysis

Note: [1] See McCool, Clark & Stankey (2007) and Manning (2011) for in-depth comparisons.

in any VUM context: (1) establishing management objectives (including types of experiences to offer), (2) identifying indicators and thresholds, (3) monitoring, (4) implementing management actions, and (5) repeating steps 3 to 5 (Figure 5.1) (Manning, 2011). These steps are detailed below.

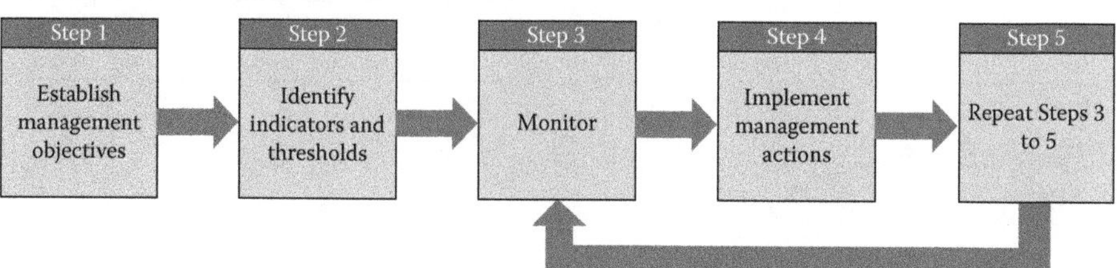

Source: Adapted from Manning (2011).

Figure 5.1 A generalized five-step Management-By-Objectives framework

Establishing management objectives

Management objectives are broad, narrative statements that describe the desired conditions of visitor experiences. Management objectives can be created using foundation documents of the tourism area, empirical research (such as REP or outcomes-based inquiry), public engagement, managerial judgment, and a variety of other components. For instance, empirical inquiry into visitor REP's may reveal that one of the primary experiences people seek in an area is solitude. The management objective for the area could be broadly written to describe what the visitor experience of solitude would be like.

Identifying indicators and thresholds

Indicators and thresholds are related concepts. Indicators are quantifiable, manageable components of the visitor experience that relate back to the management objective. It is important to emphasize that indicators related to visitor experience should be empirically developed whenever possible. A common approach to empirically developing indicators is to use an adapted important-performance evaluation (Hollenhorst & Gardner, 1994; Miller, Taff, & Newman, 2018; Pilcher, Newman & Manning, 2009). For instance, if a management objective described a component of the desired conditions as "outstanding opportunities to experience natural soundscapes," visitors can rate the importance of different soundscape indicators, such as human voices, aircraft, vehicles, birdsong and insects, among others. The indicators that have the highest valence and are relatively common would make good indicators.

Any area that accommodates tourism will see some form of impact to visitor experiences. Although indicators inform *what* components of the visitor experience to measure, thresholds help evaluate *how much change* in the indicator is acceptable. Thresholds are defined as the minimally acceptable conditions, or the point at which the visitor experience becomes unacceptable (IVUMF, 2016). Developing thresholds often involves visitor evaluations of an indicator over a range of conditions, usually through normative methods (Anderson et al., 2010; Hallo, Manning & Stokowski, 2009; Manning, 2011; Pilcher, Newman & Manning, 2009; Vaske et al., 1986). When visitor responses to indicator conditions are plotted, it results in a norm curve (Figure 5.2). The range of conditions for the indicator are plotted

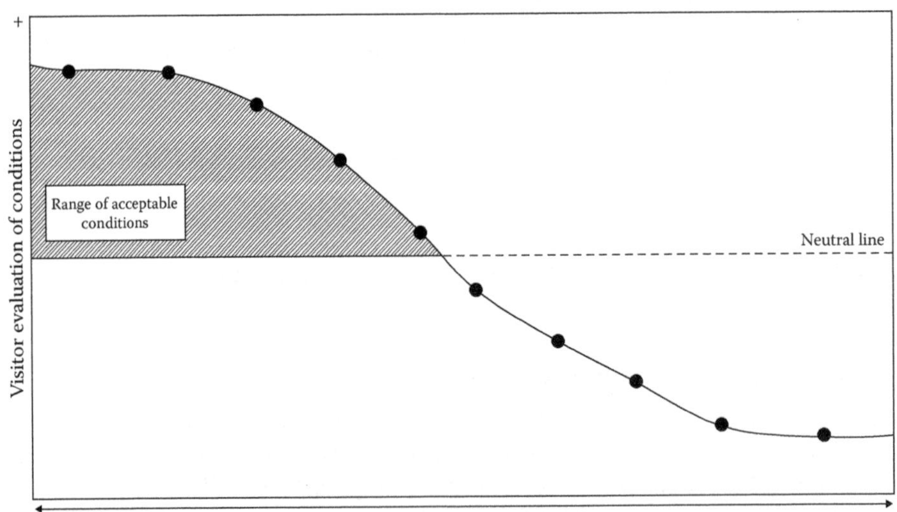

Source: Adapted from Manning (2011).

Figure 5.2 An example of a norm curve used to develop thresholds

consecutively on the x-axis, and visitor evaluations of conditions are plotted on the y-axis. Evaluation is usually measured by level of acceptability. As conditions change, visitor evaluations would also change. The point at which visitor evaluations drop below the neutral line of the plot is the threshold. The grey-shaded area above the neutral line and below the norm curve is called the range of acceptable conditions. If the indicator is within the range of acceptable conditions, visitor experience is considered satisfactory.

Monitoring

Monitoring is a process of continually generating information related to an indicator that provides critical information to sustainable tourism managers about existing conditions, changes in resource quality and the effectiveness of management actions. Monitoring takes time and dedication to do but is an essential part of the MBO process. The type and scope of monitoring is necessarily dependent on the indicators developed. For instance, monitoring might include decibel measurements, counts of visitors, average trail widths, or a variety of other measures. A detailed, repeatable, and reliable monitoring protocol is helpful for properly monitoring conditions.

Implementing management actions

Management actions are any deliberate interventions that are intended to impact the quality of the visitor experience. These include indirect management (e.g., information, education, persuasion), direct management (e.g., regulations, restrictions, enforcement), and "no action" management (no deliberate interventions are taken). These management actions should be contextual to the indicators and site where they will be implemented and can be informed through primary data collection, a review of past literature, managerial judgment or expert advice.

Repeating steps 3 to 5

After implementing actions, including "no action," the MBO framework loops back to Step 3 (Monitoring) (see Figure 5.1). In this way, managers of sustainable tourism can understand the effectiveness of their interventions. As an example (see Box 5.1), managers implemented education to reduce human-caused sound. Continued monitoring efforts demonstrated that their management actions were successful in keeping human-caused sound within the range of acceptable conditions (Figure 5.2). If monitoring shows that there is no beneficial change, additional management actions can be taken, and monitoring continued. In this way, managers can constantly learn and adapt to the complex social-ecological systems where sustainable tourism occurs.

Filling in the gaps

Although understanding and managing for visitor experiences has progressed over several decades, the complex and dynamic social-ecological contexts in which the visitor experience is embedded are still being recognized. In this section, we use a systems-thinking approach to identify challenges in the research in the form of several questions listed below. It is our hope that the inquiry related to these questions will help both research and management components of the visitor experience in sustainable tourism settings.

How can MBO frameworks be applied at different scales?

Most MBO frameworks are fairly limited in scale, ranging from sites (VIM) to areas (VERP/LAC). Often, the plans developed from these MBO frameworks fail to recognize the context of the site or area within a larger social-ecological system. This can result in unintended consequences (Box 5.2). Future research should explore how to better integrate sites into a more dynamic, large-scale system (Miller et al., 2017). This will undoubtedly be a complex and difficult task; areas have a diversity of management objectives that may be completely different, or even competing. One area of exploration related to this is the concept of *indicator places* (Miller, Freimund & Dalenberg, 2017). These indicator places would serve as proxies of conditions elsewhere in a system and would maximize the efficiency of monitoring. If conditions at an indicator place began changing in a negative direction, monitoring efforts could then be expanded to other proxy areas. Research into developing these indicator places will prove useful. Other areas of inquiry could include exploring manager perceptions of visitor use management through concepts like mental modeling (Mosimane et al., 2014) to better understand the challenges associated with undertaking system-scale management plans.

Box 5.2: In Rocky Mountain National Park (Colorado, USA), computer modeling explored how alternative transportation systems impact visitor experiences. Although there were beneficial outcomes related to visitor transportation experiences by increasing shuttle access, there were negative consequences for other components of the visitor experience. By expanding access to areas via a shuttle system, the number of visitors roughly *doubled* compared to private vehicle access only. This increase in visitor use would violate the management objectives of the area by increasing perceptions of crowding. The lesson in this study is that it is critical for managers to place their plans in the context of a greater social-ecological system (Lawson et al., 2011).

How can models for understanding thresholds be improved?

Research into the methods for developing norm curves indicates that current approaches may be flawed (Gibson et al., 2012). For instance, the order of image presentation in visual-based method studies, which are often used in visitor use management, has an effect on image evaluations. Inquiry into new methods for estimating norm curves may increase the validity and rigor of these methods. One possible approach is to use regression-based models to develop social norm curves and associated thresholds (Newman et al., 2018). Another approach is using logit models to understand the likelihood of negative impacts to visitors over a range of conditions. Whatever methods eventually emerge to overcome the challenges of traditional social norm curve methods (Gibson et al., 2012), they should be tested and compared to alternative MBO methods so that the field of visitor use management can be guided by the best available sound science.

An additional neglected concept that merits more exploration is how threshold models differ among groups. The "average camper who doesn't exist" has long been part of the narrative in planning for visitor experiences (Shafer, 1969). This concept focuses on the fact that visitors come to parks and protected areas with diverse motivations, expectations and desired outcomes that impact how they perceive the quality of their experience. However, threshold models are almost always developed for a "general population" of park visitors (for an exception, see Anderson et al., 2010). By doing this, these models fail to acknowledge the diversity of park visitors. Importantly, some of these subgroups may have characteristics that more closely align with (or oppose) the management objectives for an area, and thus would be more useful for developing threshold concepts. Additional information about how thresholds differ among types of park visitors would be helpful to managers and scientists, allowing a better understanding of how to manage for quality visitor experiences.

How can capacity for visitor use management in parks and conservation areas be built?

Park and conservation areas have excelled at building capacity related to managing ecological resources. However, the shift to viewing park and conservation areas as social-ecological systems highlights the need to build capacity for managing the social side of the equation. Historically, there has been a lack of expertise to investigate, plan for, and manage visitor use in the agencies and organizations that govern park and conservation areas (Miller et al., 2017). To overcome this barrier, it is incumbent on park and conservation area agencies to begin to seriously address visitor experience and visitor use management by providing training, shaping programs in higher education, and hiring experts to work in and address the human dimensions of park and conservation area management. Empirical research can help guide this process. For instance, what are managers' perceptions of the barriers for increasing social science capacity? What skills do they see as most important for a visitor use expert? What do managers believe is the role of social science and visitor studies? Answers to these questions may help increase the capacity for visitor use management and strengthen the resiliency of the social-ecological systems that are parks and conservation areas.

What does adaptive management mean for visitor experience?

Prior research using MBO frameworks has focused on identifying the point at which conditions become unacceptable (see Figure 5.2). However, this has been misinterpreted in relation to management; the threshold crossover is not the *only* place where managers can take action. Instead, threshold curves should be viewed as a way to understand changes in conditions, not as a static point for taking action. Informed through the threshold curves, managers can use them to identify multiple points where they would take a variety of actions depending on the severity and direction of the change required. Future research that focuses on empirically identifying the points on a norm curve at which actions should be taken would be an important part of using a science-based strategy to adaptive management.

Adaptive management in the provisioning of visitor experiences may also mean confronting and integrating the inherent complexity of such a task. For visitor use management, this includes moving from single-loop (see Figure 5.1) to double-loop learning (Argyris, 1977). Double-loop learning is a process that helps people "think more deeply about their own assumptions and beliefs" (Cartwright, 2002). Applied to MBO frameworks, the process may need additional loops that help practitioners better refine their understanding and management of visitor experiences (Figure 5.3). For instance, monitoring may find that conditions are well above or below thresholds. In this instance, managers may want to challenge their initial assumptions and beliefs about the area by adjusting their management objectives (the top arrow in Figure 5.3). This is a single example of how to incorporate double-loop learning into the management of visitor experiences. Many other routes for double-loop learning are surely possible in MBO frameworks and in the general management of visitor use, and valuable insights can be gained from thinking through the fundamental assumptions and beliefs held about the process.

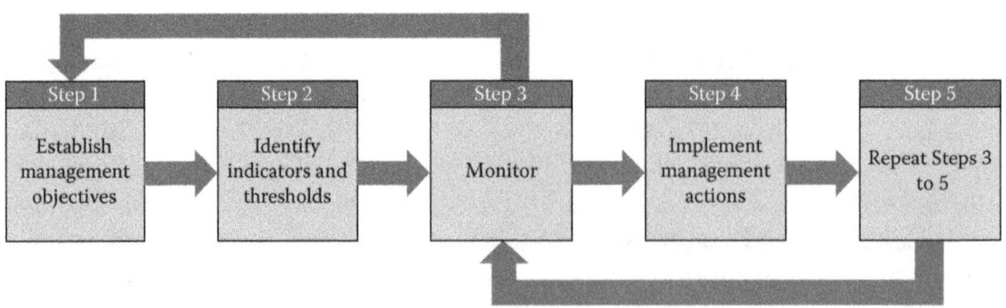

Source: Adapted from Manning (2011).

Figure 5.3 An example of double-loop learning in the generalized MBO framework

How do MBO frameworks apply in different cultural contexts?

With few exceptions (Manidis Roberts, 1997), MBO frameworks have been researched, developed, and applied by public land agencies in the USA. Currently, it is uncertain how well the concepts in MBO frameworks apply in different cultural contexts. Many basic questions, as well as all the questions raised above, need to be explored in different and cross-cultural contexts (Miller, Freimund & Blackford, 2018). For instance, the barriers and constraints for implementing MBO frameworks (mainly ROS and its adaptations) may be vastly different between the USA and other countries. Additionally, threshold concepts in MBO frameworks are usually derived from normative concepts and the idea of acceptability. In other cultures, these normative concepts may be radically different, or possibly not appropriate for developing thresholds in sustainable tourism contexts. Future applications of and research into MBO frameworks in international contexts is sorely needed.

How can sustainable tourism destinations be managed as social-ecological systems?

Protected areas characteristically have semi-porous borders and dual social-ecological missions. Managing sustainable tourism sites is as much an ecological practice as it is social. Therefore, OFM sometimes falls short of holistic management through its heavy emphasis on maximizing positive recreational benefits. It therefore makes sense to implement OFM as a component of a larger ecosystem services framework. Ecosystem services (ES) are those benefits humans obtain from the natural world. A variety of ES frameworks have been put forward to illustrate how natural resources flow into social and managerial systems to create a unit of human well-being (i.e., Costanza et al., 2014). Cultural ecosystem services (CES), are those most often associated with recreation—the nonmaterial, intangible benefits stemming from the natural world. In this way, CES and OFM-derived outcomes are strikingly similar. In the context of sustainable tourism, the linkage of their respective service/outcome domains, unique methodologies, or theory is crucial. By harnessing the broader, encompassing scale and eco-centric nature of an ES framework and the social-psychological theory and outcome domains of OFM, managers have a more holistic means of managing their sustainable tourism sites.

In a unified framework, moving from broadest to finest spatial and human scales, ecological and managerial inputs (e.g., a tidal marsh and a boat ramp) provision a set of recreational opportunities (e.g., kayaking, canoeing, paddleboarding). Based on their individual or aggregate motivations (e.g., reduced stress) and available information, visitors choose an activity from these opportunities from which they seek to derive a bundle of services (or outcomes). However, in the pursuit of these recreational services, visitors may encounter ecological or social conditions (e.g., invasive species or congestion) that might detract or enhance the quality or magnitude of their bundle of services. It follows that the achievement of a certain bundle of services informs future recreational decisions by visitors and managerial actions.

Moving forward

Since the first parks and protected areas were initially established, visitor use and experience has been an integral part of their management. Although a large body of work has developed to inform the management of visitor experiences in parks and protected areas, many challenges still remain. As parks and protected areas become increasingly important for protecting biodiversity and achieving conservation goals as well as providing humans with a variety of benefits, the proper management of visitor experiences within these parks and protected areas has never been more important.

Although the challenges identified above are a start to moving forward inquiry into the visitor experience, there are many more challenges worthy of exploration. Particularly as technology advances, there are opportunities to gain understanding from laboratory-controlled experiments, Big Data, and other emerging technologies. Additionally, much of the research conducted related to visitor experiences relies on self-reported measures—using methods of research that incorporate observational behavior, physiological responses, and quasi-experimental designs is sorely needed. Lastly, longitudinal studies need to be conducted to allow us to understand the long-term feedback loop effects of visitor experiences.

Like many other issues, sustainable tourism will need to confront the uncertainty and complexity of managing visitor experiences as we move forward. Visitor use management frameworks provide opportunities to address this uncertainty by creating institutionalized learning processes (Senge, 2006). Climate change, the relevancy of parks and protected areas to the public, shifting demographics and urbanization, technological advancements, changing political structures and myriad other factors will certainly challenge our abilities to successfully manage the visitor experience. Given these circumstances, we believe that the questions asked in the previous section can help managers of sustainable tourism challenge mental models, create positive changes in system structures, and help create authentic resolution to visitor experience challenges that we cannot yet envision.

References

Anderson, L. E., Manning, R. E., Valliere, W. A. & Hallo, J. C. (2010). "Normative standards for wildlife viewing in parks and protected areas." *Human Dimensions of Wildlife*, **15**(1): 1–15.

Ardoin, N. M., Wheaton, M., Bowers, A. W., Hunt, C. A. & Durham, W. H. (2015). "Nature-based tourism's impact on environmental knowledge, attitudes, and behavior: a review and analysis of the literature and potential future research." *Journal of Sustainable Tourism*, **23**(6): 838–58.

Argyris, C. (1977). "Double loop learning in organizations." *Harvard Business Review*, **55**(5): 115–25.

Borrie, W. T. & Roggenbuck, J. W. (1998). "Describing the wilderness experience at Juniper Prairie Wilderness using experience sampling methods." Wilderness & natural areas in eastern North America: research, management and planning. Nacogdoches, TX: Stephen F. Austin State University, Arthur Temple College of Forestry, Center for Applied Studies, pp. 165–72.

Boyd, S. W. & Butler, R. W. (1996). "Managing ecotourism: an opportunity spectrum approach." *Tourism Management*, **17**(8): 557–66.

Cartwright, S. (2002). "Double-loop learning: a concept and process for leadership educators." *Journal of Leadership Education*, **1**(1): 68–71.

Castellanos-Verdugo, M., Vega-Vázquez, M., Oviedo-García, M. Á., & Orgaz-Agüera, F. (2016). "The relevance of psychological factors in the ecotourist experience satisfaction through ecotourist site perceived value." *Journal of Cleaner Production*, **124**: 226–35.

Costanza, R., de Groot, R., Sutton, P., Van der Ploeg, S., Anderson, S. J., Kubiszewski, I. . . . & Turner, R. K. (2014). "Changes in the global value of ecosystem services." *Global Environmental Change*, **26**: 152–8.

Driver, B. L. (1976). "Quantification of outdoor recreationists' preferences." In B. Smissen & J. Myers, eds., *Research: Camping and Environmental Education, HPEP Series No. II*, pp. 165–87. University Park: Pennsylvania State University Press.

Driver, B. L. (1983). *Master List of Items for Recreation Experience Preference Scales and Domains*. Fort Collins, CO: USDA Forest Service Rocky Mountain Forest and Range Experiment Station.

Driver, B. L. (2008). "What is outcomes-focused management?" In B. L. Driver, ed., *Managing to Optimize the Beneficial Outcomes of Recreation*, pp. 19–37. State College, PA: Venture Publishing.

Driver, B. L. & Brown, P. J. (1975). "Social psychological definition of recreation demand, with implications for recreation resource planning." In *Assessing Demand for Outdoor Recreation*, pp. 65–88. Washington, DC: National Academy of Sciences.

Driver, B. L., Brown, P. J., Stankey, G. H. & Gregoire, T. J. (1987). "The ROS planning system: evolution, basic concepts and research needed." *Leisure Sciences*, **9**: 201–12.

Driver, B. L. & S. R. Tocher (1970). "Toward a behavioral interpretation of recreational engagements, with implications for planning." In B. L. Driver, ed., *Elements of Outdoor Recreation Planning*, pp. 9–31. Ann Arbor: University of Michigan Press.

Gibson, A. W., Newman, P., Lawson, S., Fristrup, K., Benfield, J. A., Bell, P. A. & Nurse, G. A. (2012). "Photograph presentation order and range effects in visual-based outdoor recreation research." *Leisure Sciences*, **36**(2): 183–205.

Graefe, A. R., Kuss, F. R. & Vaske, J. J. (1990). *Visitor Impact Management: The Planning Framework*, vol. II. Washington, DC: National Parks and Conservation Association.

Gundersen, V., Tangeland, T. & Kaltenborn, B. P. (2015). "Planning for recreation along the opportunity spectrum: the case of Oslo, Norway." *Urban Forestry & Urban Greening*, **14**: 210–17.

Haas, G. E. & Aukerman, R. (2004). "WROS: Water Recreation Opportunity Spectrum." *Lakeline*, **24**(2): 15–17.

Hallo, J. C., Manning, R. E. & Stokowski, P. A. (2009). "Understanding and managing the off-road vehicle experience: indicators of quality." *Managing Leisure*, **14**(3): 195–209.

Ham, S. (2013). *Interpretation: Making a Difference on Purpose.* Golden, CO: Fulcrum Publishing.

Han, J. H., Lee, M. J. & Hwang, Y. S. (2016). "Tourists' environmentally responsible behavior in response to climate change and tourist experiences in nature-based tourism." *Sustainability*, **8**(7): 1–14.

Hof, M. & Lime, D. W. (1997). "Visitor experience and resource protection framework in the national park system: rationale, current status, and future direction." In Stephen F. McCool and David N. Cole, compilers, *Proceedings – Limits of Acceptable Change and Related Planning Process: Progress and Future Directions*, pp. 26–9. Ogden, UT: US Department of Agriculture, Forest Service, Rocky Mountain Research Station.

Hollenhorst, S. & Gardner, L. (1994). "The indicator performance estimator approach to determining acceptable wilderness conditions." *Journal of Environmental Management*, **18**: 901–06.

IVUMF (Interagency Visitor Use Management Framework) (2016). *The Interagency Visitor Use Management Council.* Available at http://visitorusemanagement.nps.gov.

Jorgenson, J. & Nickerson, N. (2015). "The role of social science in predicting support for Yellowstone National Park." *Yellowstone Science*, **23**(1): 3–4.

Jorgenson, J., Nickerson, N., Dalenberg, D., Angle, J., Metcalf, E. & Freimund, W. (2018). "Measuring visitor experiences: creating and testing the tourism autobiographical memory scale." *Journal of Travel Research*. Available at https://doi.org/10.1177/0047287518764344.

Joyce, K. & Sutton, S. (2009). "A method for automatic generation of the Recreation Opportunity Spectrum in New Zealand." *Applied Geography*, **29**: 409–18.

Lawson, S., Chamberlin, R., Choi, J., Swanson, B., Kiser, B., Newman, P., . . . & Gamble, L. (2011). "Modeling the effects of shuttle service on transportation system performance Rocky Mountain National Park." *Transportation Research Record: Journal of the Transportation Research Board*, **2244**: 97–106.

McCool, S. F. (2006). "Managing for visitor experiences in protected areas: promising opportunities and fundamental challenges." *Parks*, **16**(2): 3–9.

McCool, S. F., Clark, R. N. & Stankey, G. (2007). "An assessment of frameworks useful for public land recreation planning." Gen. Tech Rep. PNW-GTR-705. Portland, OR: US Department of Agriculture, Forest Service, Pacific Northwest Research Station.

McCool, S. F., Freimund, W. A., Gorricho, J., Kohl, J. & Biggs, H. (2014). "Benefiting from complexity thinking." In G. L. Worboys, M. Lockwood, A. Kothari, S. Feary & I. Pulsford, eds., *Protected Area Governance and Management*, pp. 3–35. 1st edn. Canberra: ANU Press.

McCool, S. F. & Lime, D. W. (2001). "Tourism carrying capacity: tempting fantasy or useful reality?" *Journal of Sustainable Tourism*, **9**(5): 372–88.

Manfredo, M. J., Driver, B. L. & Tarrant, M. A. (1996). "Measuring leisure motivation: a meta-analysis of the recreation experience preference scales." *Journal of Leisure Research*, **28**(3): 188–213.

Manidis Roberts. (1997). *Developing a Tourism Optimisation Management Model (TOMM): A Model to Monitor and Manage Tourism on Kangaroo Island, South Australia*. Surry Hills, NSW: Manidis Roberts Consultants.

Manning, R. E. (2011). *Studies in Outdoor Recreation: Search and Research for Satisfaction*. 3rd edn. Corvallis, OR: Oregon State University Press.

Martin, S. R., Marsolais, J. & Rolloff, D. (2009). "Visitor perceptions of appropriate management actions across the Recreation Opportunity Spectrum." *Journal of Park and Recreation Administration*, **27**(1): 56–69.

Miller, Z. D., Fefer, J. P., Kraja, A., Lash, B. & Freimund, W. (2017). "Perspectives on visitor use management in the National Parks." *The George Wright Forum*, **34**(1): 37–44.

Miller, Z. D., Freimund, W. & Blackford, T. (2018). "Communication perspectives about bison safety in Yellowstone National Park: a comparison of international and North American visitors." *Journal of Park and Recreation Administration*, **36**(1): 176–86.

Miller, Z. D., Freimund, W. & Dalenberg, D. (2017). *Going-to-the-Sun: Road Monitoring Protocols*. Missoula, MT: University of Montana Department of Society and Conservation.

Miller, Z. D., Taff, B. D. & Newman, P. (2018). "Visitor experiences of wilderness soundscapes in Denali National Park and Preserve." *International Journal of Wilderness*, **24**(2): 32–43.

Ministerio de Ambiente. (2015). *Manual para la Elaboración de los Planes de Uso Público en las Áreas Protegidas del SINAP*. Panamá: Ministerio de Ambiente.

Moore, S. A., Rodger, K. & Taplin, R. (2015). "Moving beyond visitor satisfaction to loyalty in nature-based tourism: a review and research agenda." *Current Issues in Tourism*, **18**(7): 667–83.

Mosimane, A. W., McCool, S., Brown, P. & Ingrebretson, J. (2014). "Using mental models in the analysis of human–wildlife conflict from the perspective of a social–ecological system in Namibia." *Oryx*, **48**(1): 64–70.

Newman, P., Taff, B. D., Ferguson, L. A., Graf, A., Costigan, H. & Miller, Z. D. (2018). *Denali National Park and Preserve Frontcountry Sounds: Impacts of Sounds on Visitors' Experience in Denali National Park and Preserve's Frontcountry*. University Park, PA: Pennsylvania State University, Department of Recreation, Park, and Tourism Management.

Organic Act of 1916. 16 U.S.C. §§ 1–4.

Pilcher, E. J., Newman, P. & Manning, R. E. (2009). "Understanding and managing experiential aspects of soundscapes at Muir Woods National Monument." *Environmental Management*, **43**: 425–35.

Pinkus, E., Moore, S. A., Taplin, R. & Pearce, J. (2016). "Re-thinking visitor loyalty at 'once in a lifetime' nature-based tourism destinations: empirical evidence from Purnululu National Park, Australia." *Journal of Outdoor Recreation and Tourism*, **16**: 7–15.

Powell, R. B. & Ham, S. H. (2008). "Can ecotourism interpretation really lead to pro-conservation knowledge, attitudes and behaviour? Evidence from the Galapagos Islands." *Journal of Sustainable Tourism*, **16**(4): 467.

Romagosa, F., Eagles, P. F. & Duitschaever, W. B. (2012). "Evaluación de la gobernanza en los espacios naturales protegidos. El caso de la Columbia Británica y Ontario (Canadá) 1." In *Anales de Geografía de la Universidad Complutense* (vol. 32, no. 1, p. 133). Universidad Complutense de Madrid.

Senge, P. (2006). *The Fifth Discipline: The Art and Practice of the Learning Organization*. New York: Doubleday/Currency.

Shafer, E. L. (1969). "The average camper who doesn't exist." Res. Pap. NE-142, p. 142. Upper Darby, PA: US Department of Agriculture, Forest Service, Northeastern Forest Experiment Station.

Snyman, S. (2017). "The role of private sector ecotourism in local socio-economic development in southern Africa." *Journal of Ecotourism*, **16**(3): 247–68.

Stack, D. W., Newman, P., Manning, R. E. & Fristrup, K. M. (2011). "Reducing visitor noise levels at Muir Woods National Monument using experimental management." *The Journal of the Acoustical Society of America*, **129**(3): 1375–80.

Stankey, G. H. (1977). "Some social concepts for outdoor recreation planning." In *Outdoor Recreation: Advances in Application of Economics*, pp. 154–61. General Technical Report WO-2. Washington, DC: USDA Forest Service.

Stankey, G. H., Cole, D. N., Lucas, R. C., Peterson, M. E. & Frissell, S. S. (1985). *The Limits of Acceptable Change (LAC) System for Wilderness Planning*. Ogden, UT: US Department of Agriculture, Forest Service, Intermountain Forest and Range Experiment Station.

Stankey, G. H., McCool, S. F. & Stokes, G. L. (1984). "Limits of acceptable change: a new framework for managing the Bob Marshall Wilderness Complex." *Western Wildlands*, **10**(3): 33–7.

Sumner, E. L. (1936). *Special Report on a Wildlife Study in the High Sierra in Sequoia and Yosemite National Parks and Adjacent Territory*. Washington, DC: US National Park Service Records, National Archives.

Valdivieso, J. C., Eagles, P. F. & Gil, J. C. (2015). "Efficient management capacity evaluation of tourism in protected areas." *Journal of Environmental Planning and Management*, **58**(9): 1544–61.

Vaske, J. J., Shelby, B., Graefe, A. R. & Heberlein, T. A. (1986). "Backcountry encounter norms: theory, method, and empirical evidence." *Journal of Leisure Research*, **18**(3): 113–38.

Wagar, J. A. (1964). *The Carrying Capacity of Wild Lands for Recreation*. Washington, DC: Society of American Foresters.

Whittaker, D., Shelby, B., Manning, R. E., Cole, D. & Haas, G. (2011). "Capacity reconsidered: finding consensus and clarifying differences." *Journal of Park and Recreation Administration*, **29**(1): 1–20.

Yellowstone Park Act of 1872. 16 U.S.C. §§ 21.

Yosemite Grant Act of 1864. 16 U.S.C. §§ 48.

6 Connecting people with experiences

Gianna Moscardo

Introduction and context

The experiential turn in tourism research explicitly recognizes that the central element driving tourist behavior is the desire to have a positive experience at a destination (Moscardo, 2015a). Tourist destinations offer tourist experience opportunities through their physical features, the activities offered and the support infrastructure provided. Traditionally destination marketers and tourism developers focused on identifying the most readily accessible and most easily tempted tourist groups or markets and adjusting the destination to offer the experience opportunities judged to be most attractive to these markets. This approach, referred to as promotional-dominated marketing by McCool in Chapter 1 of this book, can be linked to a wide variety of unsustainable practices and outcomes.

Despite decades of discussion about sustainability in the tourism literature, global tourism is arguably less sustainable than it has ever been and there is little systematic evidence that tourism is an effective element of sustainable development (Buckley, 2012; Hall, 2010). Recent reviews of tourism's sustainability are not encouraging, with reports of significant and increasing contributions to climate change (Lenzen et al., 2018), concerns over excessive use of resources (Gössling & Peeters, 2015), problems with congestion and overcrowding (Jordan & Spencer, 2017) and displacement of destination residents (Zanini, 2017). The phrase "overtourism" has been used in the news media as a catchall phrase for situations where the extent and significance of the negative impacts of tourism outweigh the positive benefits resulting in destination resident backlash against tourism. Although this phrase is already the subject of considerable academic angst about definitional issues with a focus on crowding and carrying capacity, more in-depth analysis of this backlash indicates that it is driven by many issues, not just tourist numbers or crowds (Alexis, 2017; Zanini, 2017). This media coverage of overtourism is actually about the consequences of a failure to implement sustainability in tourism practice. Arguably this is also a failure of tourism academics to influence tourism practice in any significant way.

McCool (this volume, Chapter 1) suggests that one reason for this failure of tourism research to make any significant difference to the sustainability of tourism in practice has been a reliance on traditional descriptive frameworks and simplistic linear causal models of tourism, and a failure to fully embrace systems thinking. These concerns have also been raised by others in tourism (Bramwell et al., 2017; Moscardo, 2017a; Ruhanen, Moyle & Moyle, 2018). These themes are also apparent in the wider planning and development literature where concerns have been raised about the effectiveness of what have been called technical, rationalist, positivist approaches that are based on unsupported assumptions about linear cause-and-effect relationships (Brown, 2012; Green & Haines, 2012; Moscardo, 2019). Similar arguments exist in the broader sustainability literature (e.g., Bai et al., 2016).

Given these challenges and the need to engage in more complex systems thinking, this chapter uses experience as a concept that connects tourists to destinations, and explores how these connections are linked to tourism impacts and sustainability. This systems approach is used to map out the different ways tourists and residents are connected to each other and to tourist experience opportunities. It has been argued that while this tourism system operates in a world of rapid change, existing market driven processes in tourism that identify and connect tourists to the potential experiences offered at a destination have changed very little. It is this slowness to adopt new approaches that directly threatens the ability of tourism to contribute in any significant way to sustainability.

The present chapter makes an important distinction between tourist experiences and experience opportunities. Tourist experiences can be defined as memorable episodes "within the constant stream of activity and sensory input that make up human lives, that occur within a specific time period and spatial context, and that are associated with emotional responses, personal meaning and significant memories" (Moscardo, 2017b, p. 98). Tourist experiences can only be created, remembered and reported by the individual tourist and those responsible for tourism practice at a destination can only offer the potential for tourist experiences. Destination developers, marketers, managers and providers offer tourist experience opportunities and it is these experience opportunities that are promoted in marketing to encourage tourists to visit a destination. The focus in the present chapter is on the way destinations create, promote and manage these experience opportunities and how this influences the sustainability of tourism at the destination level.

This chapter adopts a destination community well-being (DCW) approach to sustainability which argues that several different forms of capital including natural, social, human, cultural, political, built, and manufactured as well as financial capital, are critical elements of well-being for both individuals and communities (Bandarage, 2013; Scott, 2012). This DCW approach to sustainability also proposes that new models of economic systems need to explicitly include all these capitals and focus on improving overall well-being and not just financial returns if they are to support greater sustainability (Costanza et al., 2010). Based on this broader approach to sustainability the chapter uses Moscardo et al.'s (2017) definition of tourism sustainability "as tourism activities that maintain and enhance destination community well-being through net contributions to all forms of capital, especially natural capital" (Moscardo et al., 2017, p. 286). The core argument here is that tourism can have both positive and negative impacts on a range of different capitals that make up overall DCW. Understanding and managing tourism impacts on these DCW capitals is therefore critical to make tourism sustainable.

Systems thinking

This chapter takes an epistemological perspective which sees a systems approach as a way of thinking about and analyzing situations, rather than an ontological view which seeks to describe processes in detail (Abson et al., 2017). Thus, systems thinking can be defined as "a set of synergistic analytic skills used to improve the capability of identifying and understanding systems, predicting their behaviors, and devising modifications to them in order to produce desired effects" (Arnold & Wade, 2015, p. 675). Systems thinking has three essential characteristics—it is holistic rather than reductionist, it seeks to improve interventions aimed at reaching desired futures, and it simplifies complex situations through the use of conceptual models (Arnold & Wade, 2015). Systems thinking focuses more on the whole situation of interest rather than on its elements, and on how the elements interact to produce change, rather than on describing the details of the elements (Bai et al., 2016; McCool, Freimund & Breen, 2015; Merali & Allen, 2011; Mingers & White, 2010). Of particular importance are emergent properties which are characteristics that can only be discerned from an examination of the functioning whole (Abson et al., 2017; Merali & Allen, 2011). It is also important in systems thinking to consider the purpose of both the system analysis and the system itself (Hodgson & Midgley, 2014; McCool, Freimund & Breen, 2015).

Finally, systems thinking progresses from analysis to action through simplifying and representing the issue or problem in a conceptual model (Arnold & Wade, 2015; McCool, Freimund & Breen, 2015; Merali & Allen, 2011). These systems have the following features:

- Elements, which are major characteristics of the system including key actors or institutions (Arnold & Wade, 2015);
- Interactions, interconnections and relationships between these elements (Arnold & Wade, 2015; Mingers & White, 2010);
- Feedback loops connecting outcomes or consequences back to elements (McCool, Freimund & Breen, 2015); and
- Leverage points where intervention can change the system and its outcomes (Abson et al., 2017; McCool, Freimund & Breen, 2015).

The tourist experience opportunity system

Figure 6.1 provides a conceptual systems model of tourist experience opportunity that seeks to map out the ways in which different groups of people are connected to tourist experiences. It also presents the processes that contribute to unsustainable

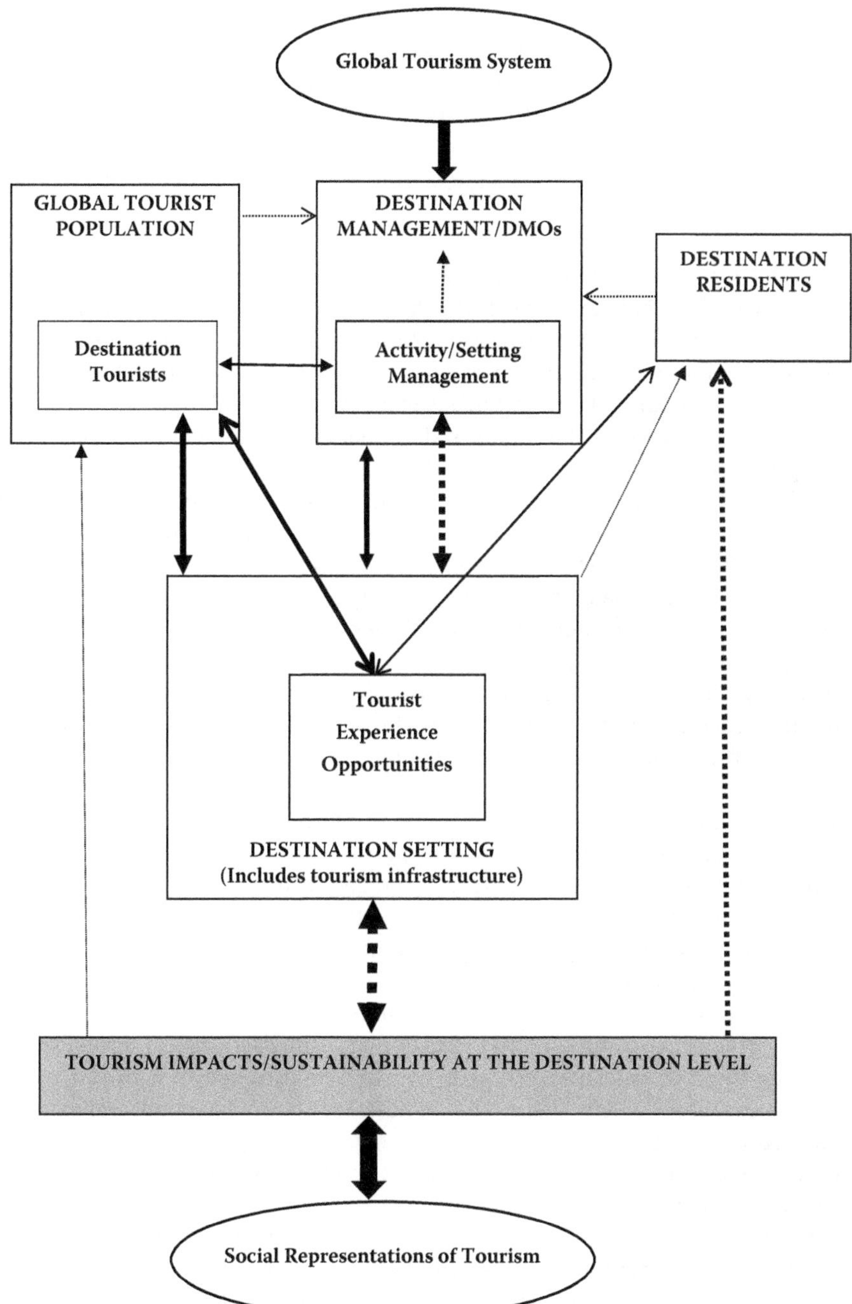

Figure 6.1 Conceptual systems model of tourist experience opportunities

outcomes for tourism destinations. The systems model identifies seven major elements that can be linked to tourist experience opportunities and these are contained in the boxes shown. Major interconnections between these elements are included as solid black arrowed lines and feedback loops are presented as dotted arrowed lines. Two major forces on the system are also included in the ovals and the core emergent property of the system, tourism impacts, are evident in the gray box.

The model can be explained using two contrasting examples of tourist experience opportunities that demonstrate the different ways in which tourist experience opportunity marketing and management can contribute to either improved destination sustainability or negative impacts on destination resident well-being—stag parties conducted in numerous European cities and climbing Uluru (formerly known as Ayers Rock). At the center of the model is the element of tourist experience opportunities which are based on the activities tourists participate in while at the destination. Stag parties are celebrations in Anglo-European wedding traditions allowing the groom one last opportunity to engage in various activities linked to being single, usually involving alcohol and substance abuse, sexual encounters and extreme physical challenges and games (Briggs & Ellis, 2017). Uluru is an iconic symbol of Australia and a popular, although remote, destination for approximately 270,000 domestic and international visitors each year. Uluru is a single red sandstone monolith rising to a height of 348 meters above the surrounding desert with a circumference of 9.4 kilometers. One of the activities that tourists can currently engage in is the climb to the top of the monolith itself.

At the top of the model in Figure 6.1 is one of the two main forces that influence the process—global tourism systems. Stag parties have emerged largely because of aspects of this global system. The rise of budget airlines and generally increased spending on weddings in the UK has seen a rise in these parties being the focus of short trips to various European cities (Iwanicki, Dluzewska & Kay, 2016). This phenomenon did not arise as the result of any specific promotional campaign on the part of a destination marketing or management organization (DMO), although a number of tour companies, mostly based outside the destinations, specializing in organizing these events have emerged. Tourist activities unfold within the larger destination setting element which has physical, economic, socio-cultural and tourism structural characteristics. Stag parties are usually conducted in bars and clubs in urban areas. The setting characteristics and nature of the tourist activity may be influenced by the third element—experience/setting management institutions. The management of tourist activities is embedded within the fourth element of destination management (Figure 6.1). In the case of stag parties, while most of the cities visited will have a DMO, the primary promotional and organizational groups are the external companies that specialize in this style of tourism. Stag parties are typical of many tourist activities in that they occur across a range of different locations within a destination with no single institution responsible for their overall management or control. While the group remains within a bar, hotel or resort they are potentially being managed by the staff of those establishments, although these private businesses may not have actively sought this type of business or be aware that this activity was going to take place on their premises. When the group is outside of these private establishments they are bound by relevant local laws, but they are typically not being directly managed by anyone. Destination settings are also often places where destination residents live and work, with residents being the fifth main element to consider. This is especially the case for the city destinations for stag parties, where residents are very likely to encounter these tourists while moving through cities for work and leisure.

Destination tourists are the sixth element of this systems model (Figure 6.1). A number of tourist characteristics influence both how they behave in the destination setting and their personal tourist experiences. These characteristics include expectations; familiarity with the setting; desired benefits; social groups including the people immediately with them, following them through social media, and their normative reference groups; personality, values and attitudes; and physical abilities. Stag party participants are younger British males focused on the activities of drinking, partying and physical challenges with the destination merely a backdrop to the desired activities (Briggs & Ellis, 2017; Thurnell-Read, 2011). An important aspect of this element is interactions with other tourists. A common feature reported for stag parties is their negative impact on other tourists in the destination (Golby, 2018; Read, 2018). This can alter the larger tourist population as some visitors shift their attention to other destinations to avoid this type of tourism.

The key emergent property of this system (Figure 6.1) is the impact of tourist experiences on the destination. These outcomes include the impacts, both positive and negative, that tourist activities have on the tourists themselves, the businesses and agencies that provide the experience opportunities, the destination residents, and the physical environment both at and beyond the destination setting. It is these impacts that determine the sustainability of tourism. These impacts can occur on two levels—the immediate impacts of the tourist engaging in the target activity and the wider impacts of having tourists in the destination overall, and through two pathways—the direct actions of the tourists and the more indirect consequences of the infrastructure that is developed to support tourism. Stag parties have been mainly linked to direct and immediate negative social, cultural and physical environmental impacts including public drunkenness, vandalism and general antisocial behavior (Golby, 2018; Gunnarsdottir, n.d.; Read, 2018; Thurnell-Read, 2011). There is little evidence available to support the existence of significant indirect, wider positive economic impacts to balance these social and environmental problems. Arguably, stag parties discourage other types of tourists from visiting and so may be indirectly contributing to a loss of economic benefits for the destination overall. Feedback loops (dotted lines in Figure 6.1), indicate that immediate and direct tourist impacts are most clearly evident in the destination setting where they occur. In the case of stag parties, vandalism, abuse, and antisocial behavior are obvious negative impacts and given that they often occur in places frequented by destination residents there is a direct feedback loop to these residents. DMOs typically, however, only receive information about the negative impacts of stag party tourism through resident and activity/setting management and if the interconnections between these three groups are weak, this feedback can be delayed. Further, if other tourists encounter stag party tourism and are unhappy with their experiences, this can alter the destination's attractiveness to the global tourism population. Again, while this is a form of feedback about tourism impacts for DMOs, it is likely to be even more delayed than feedback through residents and activity management. These indirect and slow feedback loops in tourism mean that DMO responses to sustainability issues with tourism are likely to be very slow and possibly too late.

The Uluru example is a very different one that serves to highlight a contrasting way in which the overall system can operate when interconnections between DMOs, destination residents and activity/setting management are stronger and closer. Uluru lies within a designated protected area that is closely controlled and directly managed jointly by Parks Australia, a government agency, and a management board made up of traditional owners and other destination stakeholders. The adjacent visitor services center, Yulara, was developed specifically to support tourism and is owned and managed by Voyages Indigenous Tourism Australia. This organization also acts as the main DMO for this destination, although the destination is also promoted through the regional DMO, Tourism Northern Territory. The larger region is home to the Anangu indigenous people who are therefore the main destination resident group of concern. Visitor surveys suggest that visitors to Uluru fall into three demographic groups: older Australian resident retired couples, younger European couples and middle-aged Americans traveling with family (Tourism NT, 2011). All visitors who participated in the survey had high levels of interest in cultural and nature-based travel in general and came to Uluru because it was an iconic Australian destination that offered the opportunity to learn about Aboriginal culture and the natural environment, with some also interested in adventure (Tourism NT, 2011). Overall this was a specialized, experienced group of tourists who had a specific interest in the destination itself.

In the case of the Uluru climb the immediate impacts are negative cultural impacts for the indigenous residents, negative environmental impacts on the monolith itself and negative economic impacts in terms of the resources needed to rescue tourists and manage injury and death. Balanced against these negative impacts are the issues of providing a positive tourist experience of Uluru that supports their continued visitation, which in turn provides significant positive economic and social benefits for the destination in terms of cultural support, jobs and business opportunities. The Uluru climb was identified as a problem through a tourism governance group with a clear and central role for destination residents, and focused on ways to make tourism work for environmental conservation and to improve the well-being of the destination residents. This group then worked over an extended period of time to create more sustainable alternative tourist experience opportunities and to run a tourist education program explaining why the climb was inappropriate. In order to address this balance the Uluru climb will be closed in 2019, a decision that has been delayed over time to allow for the development of alternative ways to experience the icon and maintain visitor satisfaction. This strategy has been successful with fewer and fewer visitors choosing to climb and a large percentage reporting that the destination is still attractive despite the closure (Tourism NT, 2011). In the Uluru example there are very strong linkages between destination residents and the destination marketing and management, the destination management and activity/setting management, the tourists and the activity/setting management, and between tourists and the destination residents. These strong linkages can be seen as contributing to the development and promotion of experience opportunities that are consistent with both the expectations of the visitors and the aspirations of the destination residents and that appear to be creating a sustainable balance between negative and positive tourism impacts. As a consequence of these factors, the climb closure has been a relatively conflict-free action.

In addition to the stronger interconnections between key actors in the system, the Uluru example reflects a different social representation (SR) of tourism, the second major force on the system (bottom oval in Figure 6.1). SRs can be defined as mental constructs created through social interaction that assist people to construct and navigate social reality (Moscardo, 2009; Sharpley, 2014). These SRs of tourism include beliefs about the value of tourism, socially desirable tourist experiences and actions, as well as the causes and consequences of various tourism impacts and how they vary amongst the different actors. At Uluru, tourism is seen a tool to be used by the destination residents to achieve goals beyond tourism, while stag party tourism is seen by the tourists as an opportunity to engage in egocentrism and consumption and by the commercial sector attached to the activity as an opportunity for commercial gain. In this second SR, the tourism system exists to pursue tourist and tourism business benefits only and destinations exist only as a resource for that process.

Identifying and analyzing key leverage points

Abson et al. (2017) argue that leverage points are most likely to be found by considering the role of institutions, the flows of information through interconnections in the system and people's perceptions of the purpose and nature of the system. The contrast between the Uluru climb and the stag party examples suggests that leverage points for tourism changes could exist in:

- the connections between DMOs and destination residents;
- the way tourist experience opportunities are developed, promoted and managed, especially in terms of their impacts on resident well-being; and
- the prevailing SRs of tourism.

Social representations of tourism

Crick (1989) was one of the first to argue that academics made decisions about what and how to research, based on their SRs of what tourism was and how it worked. He argued that there were two different representations of tourism in the existing academic literature—"tourism as a godsend and tourism as evil" (p. 308).

This tension between two very different tourism SRs was also evident in Jafari's (1990) discussion of different tourism platforms. Jafari argued that there were four potential approaches to tourism research:

- An advocacy platform which matched Crick's "tourism as godsend" approach and which was characterized by research measuring the benefits, especially the economic ones, of tourism;
- A cautionary platform matched to Crick's "tourism as evil" representation in which researchers focus on the negative consequences of tourism;
- An adaptancy platform where the advocacy approach is adapted to incorporate awareness of the negative consequences highlighted in the cautionary platform and the focus of tourism research is on alternative forms of, and approaches to, tourism that minimize the negative outcomes; and
- A knowledge-based platform that was presented as the goal for tourism researchers with a focus on scientific approaches to tourism that should avoid personal values and assumptions about tourism.

The key issue for the present discussion is that most tourism research still lies mainly in the adaptancy platform. Moscardo (2009) argued that most tourism researchers assume tourism can be good under the right circumstances and seem reluctant to challenge the dominant business and government approaches to tourism built around growth and the commercial success of tourism businesses. This could be driven in part by a desire to maintain positive relationships with the sector that provides support for research, validation for the existence of tourism studies and employment for graduating students, an argument not dissimilar to that offered by Tribe (1997). It could also be driven by personal interests in travel amongst tourism academics and a desire to present the research they do in a positive light (Moscardo, 2009). The problem with an advocacy-/adaptancy-based tourism SR is that it privileges the tourism commercial sector and what Tribe (1997) called the business of tourism research over other critical issues and important stakeholders.

Moscardo's (2011) review of tourism planning models presented in the academic and government literature reveals the consequences of placing the business of tourism at the center of research and practice. Her conclusions were consistent with those made by Getz in 1986 after a similar review, indicating little change in tourism planning in a thirty-year time span. These conclusions included:

- The assumption that tourism is a necessary and desirable activity with no evidence of any comparative evaluation against other development options;
- A view that tourism planning is a variation of business planning and that the goal of tourism planning is to meet market demand and/or customer needs;
- An almost exclusive focus on economic issues with some concerns expressed about environmental problems, but virtually no discussion of other sustainability dimensions;
- Greatest attention and importance are given to tourists, external agents and tourism businesses in the formulation of plans; and

- Little, if any, attention is paid to the involvement of destination residents (Getz, 1986; Moscardo, 2011).

If we accept from the point of view of the destination that tourism is a strategy to improve community well-being, then this type of planning is unlikely to achieve sustainability. This tourism SR means that the planning and management of tourism very quickly becomes about growth in tourism numbers, the success of tourism businesses and meeting market demands, rather than about how well tourism is achieving DCW goals. As Moscardo (2008) and Moscardo and Murphy (2014) note, this means that destination communities go from being the ones who should benefit from tourism to being resources to benefit tourism businesses and tourists, and tourism becomes the end in itself rather than a means to an end. The system presented in Figure 6.1 is driven by an SR of tourism that typically focuses the attention of DMOs on the global tourism population and destination tourists rather than on destination residents and the management of tourists in the destination. This advocacy-based tourism SR also means that most monitoring systems or feedback loops that are in place for tourism are about tourist numbers and satisfaction, with very few destinations measuring any other aspects of tourism or its impacts on the destination (Mihalic, 2016). Finally, this simple positive view of tourism as an activity spills over into the SR of tourism held by tourists, in which it is seen as a socially desirable way to engage in discretionary, conspicuous consumption. Moscardo's (2009) analysis of people who could but do not engage in tourism reveals a strong social pressure on people not to admit that tourism itself may not always be a positive investment of time, money or effort. Thus, there is social pressure on those with the money and ability to travel to engage in tourist activities, which in turn puts considerable pressure on the entire system.

Overall, there has been very little critical reflection on the assumptions various groups, including government planners and policymakers, DMO staff and researchers make about tourism. Tourism research is also limited by the problem of survivor bias. The survivor bias is one where conclusions are based on a sample taken from a larger population that has in some way survived or been successful and the rest of the population is ignored (Shermer, 2014). In tourism, research has focused almost exclusively on satisfied tourists, with conclusions about the benefits of tourism for tourists based solely on those that actually enjoyed their tourism experience. There is virtually no research into those that decide not to travel or who had negative experiences, or of tourism ventures that fail, giving a very incomplete picture of the true value of travel. Without research that seeks to overcome this method bias, the tourism field is likely to continue to overestimate the positive outcomes of tourism.

The role of destination residents in tourism governance

A review of one hundred case studies describing tourism development identified a series of barriers to sustainability that were consistent with the tourism SR discussed in the previous section and that highlighted a range of problems linked to limited involvement of destination residents in tourism governance

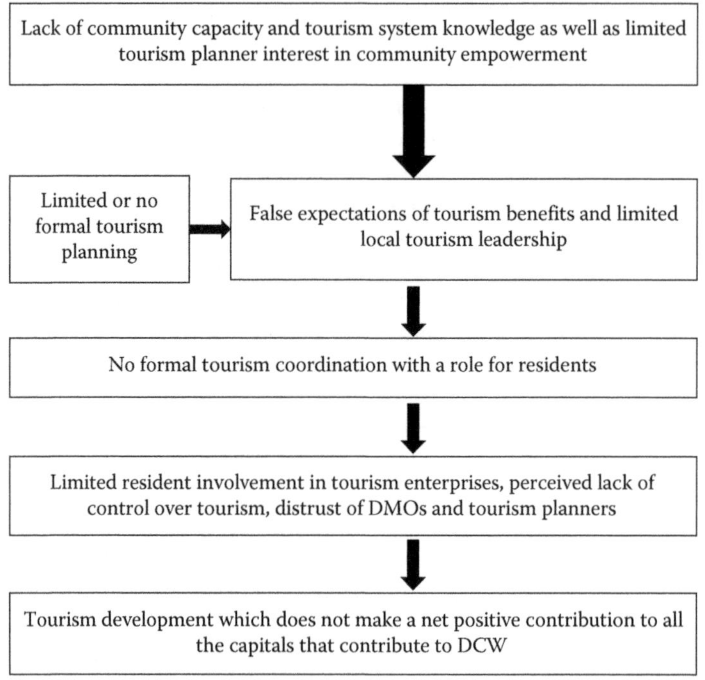

Source: Adapted from Moscardo (2011) and (2015b).

Figure 6.2 Barriers to sustainable tourism development

(Moscardo, 2011). Figure 6.2 provides a summary of these, highlighting those that are connected to the role of destination residents in tourism, including a lack of local tourism leadership which allows for greater control taken by external agents; community conflict over development proposals; limited community involvement in actual tourism businesses and activities; and a lack of perceived control over tourism leading to a lack of trust in DMOs and negative attitudes towards tourism. Although virtually all textbooks, government guidelines and academic papers argue for the involvement of the general public or community in tourism governance, it is clear that this rarely happens in actual practice (Butcher, 2012; Hewlett & Edwards, 2013; Marzuki & Hay, 2013). The existing tourism research on this topic falls into two main categories. Included in the first are studies that describe either specific case studies of public participation, typically in more peripheral destinations (cf. Bello, Lovelock & Carr, 2017) or for specialist tourism (cf. Rasoolimanesh et al., 2017). The second category includes studies that describe barriers to public involvement. Research in this latter category typically identifies community disinterest and apathy as major barriers to public engagement with tourism governance (Bello, Lovelock & Carr, 2017; Mak, Cheung & Hui, 2017; Saufi, O'Brien & Wilkins, 2014). In the wider literature on planning, this public disinterest and apathy is increasingly seen as a problem of distrust of planners and planning processes that do not allow for meaningful involvement (Baker, Hincks & Sherriff, 2010; Swapan, 2016).

The wider literature on planning and governance has attempted to address these barriers and encourage greater community empowerment in governance through:

- A focus on directly linking planning and development proposals to community well-being (Green & Haines, 2012);
- Greater time and resource investment in building community capacity and skills to support engagement in governance (Oliver & Pitt, 2013);
- Increasing attention paid to developing and supporting local social networks and community-based organizations to represent communities, especially marginalized groups, in planning and development decisions (Fisher & Shragge, 2017; Howard & Wheeler, 2015);
- The use of a wider range of more innovative public participation techniques (Glackin & Dionisio, 2016); and
- A move from more technical, positivist to post-positivist, ethical and value-based approaches to planning that explicitly seek to give power to communities (Green & Haines, 2012; Moscardo, 2019).

Despite the availability of considerable research evidence and practice examples from the broader planning and development issues demonstrating that it is both possible and desirable to give destination communities greater involvement in, and power over, tourism, there has been little change in either tourism practice or research in the last 30 years in this area. This means that feedback loops to the DMOs described in Figure 6.1 that provide information on the impacts of tourist experiences opportunities are weak.

Figure 6.3 provides an alternative approach to tourism planning that places destination communities in the center and incorporates elements from the wider contemporary literature on more effective approaches to planning for sustainability. This alternative approach gives destination residents a central role and connection to all aspects of destination tourism planning, is driven by a focus on improving DCW and suggests the use of sustainability monitoring that measures a wide range of the different dimensions or capitals that make up DCW. This latter change would improve the feedback loops in the system described in Figure 6.1.

Tourist experience opportunity development and management

The prevailing SR of tourism and its expression in the use of 1980s strategic business planning models means that DMOs have typically directed their attention to increasing tourist numbers through promotion and advertising; seen their primary stakeholders as tourism businesses; and conducted activities to develop destinations and destination communities as resources for these tourism businesses. The approach outlined in Figure 6.3 suggests, however, that DMOs should direct their attention to increasing all the capitals that contribute to DCW; see destination residents as their primary stakeholder; and shift their focus from promotion to building community capacity for tourism governance, identifying sustainable tourist markets and coordinating the development of sustainable tourist experience

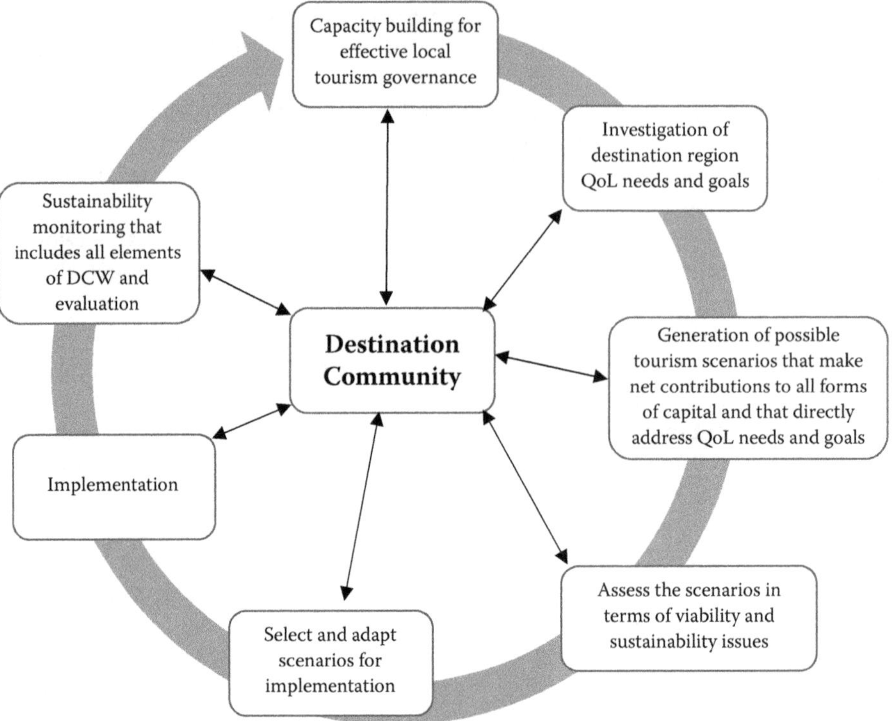

Source: Adapted from Moscardo and Murphy (2014).

Figure 6.3 An alternative approach to tourism planning for destination sustainability

opportunities. It is these last two activities, identifying sustainable tourism markets and developing experience opportunities, that are particularly relevant to the present discussion. It is timely for such a discussion of changing DMOs rules, not just because of increasing concerns over tourism sustainability, but also because there is evidence that the rise of social media and platforms such as TripAdvisor have fundamentally changed the way people look for, decide, plan and book their travel, making many of the current DMO activities obsolete (Buhalis & Amaranggana, 2014). As in the previous section, the wider literature on marketing and sustainability offers directions on how to change tourism practice and research in this area. Also as in the previous section, there is limited tourism research into these topics, with virtually no research into DMOs and their role in destination sustainability (Dwyer, 2018) and virtually no discussion of sustainability marketing in tourism marketing research (Hanna et al., 2018).

In 2011, Belz and Peattie outlined a sustainability marketing approach that offered an alternative way to conceptualize each of the major elements of traditional marketing strategy and these are summarized in Table 6.1.The stag party example is one that reflects the traditional marketing approach. Tour companies used a new

Table 6.1 Contrasting traditional and sustainability marketing elements

	Traditional marketing	Sustainability marketing
Product development	Driven by keeping production costs low and creating socially desirable features that contribute to immediate customer enjoyment	Driven by more than just economic objectives, but also ecological, social and sustainability objectives Guided by a desire to contribute to long-term customer well-being Designed to minimize negative impacts at all stages of the total product life cycle
Techniques	Greater use of: – advertising – sales promotions – direct sales – public relations	Greater use of: – information & education – public relations – external assessment & eco-labels – sustainability reporting – social marketing campaigns
Audiences	Most attention paid to: – consumers – investors	Equal attention paid to: – consumers – employees – investors – communities – competitors
Customer communication	**Awareness** Knowledge that the product exists and what its features & benefits are	**Awareness** Knowledge of product, but also its sustainability features and how it can be used and disposed of in a sustainable fashion Information about the sustainability issues relevant to the product and its use, and the business operations Programs to assist all stakeholders to understand these issues and how their actions can be linked to these issues
	Acceptance Create motivation to buy the product Focus on social desirability of consumption of the product	**Acceptance** Reporting on sustainability performance and using eco-labels to encourage trust in sustainability claims Advertising, education and social marketing to encourage consumers to believe that the issues are relevant and important and that their actions can make a difference to the sustainability issue Focus on the message of social desirability of more sustainable consumption

Table 6.1 (continued)

	Traditional marketing	Sustainability marketing
	Ability Provide information on how to access and possibly on use Use sales promotions such as discounts to encourage purchase **Action** Remind consumers of opportunities to purchase	**Ability** Education and information to support sustainable use of the product **Action** Reminders and support for engaging in sustainable use of the product Recognition and reward programs for sustainable actions for all stakeholders

Source: Belz & Peattie (2011); Moscardo (2013).

opportunity, the rise of budget airlines, to encourage a target market, young UK males, to consume a product with the main aim being economic benefits for the tour companies and suppliers. Arguably, the product provides few, if any, long-term benefits to either the consumer or the destination. The activities developed as alternatives to the Uluru climb offer an example closer to the sustainability marketing approach. One alternative is an indigenous guided tour of the base of Uluru. This tourist experience opportunity uses the cultural and natural resources of the destination to encourage an appropriate target market, with an interest in learning about culture and the setting, to consume a product that offers a range of immediate and longer term benefits for the consumer and destination.

Although Jamrozy (2007) and Pomering, Noble and Johnson (2011) have suggested the need for a marketing paradigm to support more sustainable tourism, to date the majority of tourism market research on tourist marketing has been conducted within a traditional marketing framework. Only two streams of research using sustainability to segment and identify tourist markets for a destination appear to have been published. Moscardo and Murphy (2016) focused on an island destination and segmented tourists based on their interest in sustainable tourism activities and their support for destination sustainability initiatives. The study identified and profiled a group of tourists who supported more sustainable tourism experience opportunities and organized their travel in ways that were more likely to increase DCW. Sanchez-Fernandez, Iniesta-Bonille and Cervera-Taulet (2018) segmented tourists across the Mediterranean on their interest in destinations that implement sustainability strategies in a number of different areas.

In both these studies a major challenge was identifying the features of tourist activities that could be considered as contributing to positive outcomes or sustainability.

Moscardo and Murphy (2016), for example, argued that tourists who stayed in locally owned, self-catering accommodation were more likely to contribute to local business earnings and that tourists who traveled shorter distances to the destination had a lower negative environmental impact. While there exists considerable research into the economic and environmental impacts of tourism there is much less information available on its social and cultural impacts (Deery, Jago & Fredline, 2012; Sharpley, 2014). This lack of understanding of the processes that result in social and cultural impacts makes it difficult to develop more sustainable tourist experience opportunities. However, it seems that more sustainable tourist activities are those that have a minimal ecological footprint; that involve destination residents as owners and co-participants; that support the restoration and conservation of cultural and natural heritage; that provide direct support for the development of services and facilities that are useful for destination residents; and that encourage learning for both tourists and others (Rodriguez-Diaz & Rodriguez, 2016).

Conclusions and future directions

This chapter has taken a systems thinking approach to the way DMOs create, promote and manage the experience opportunities offered to tourists in destinations, and how these actions and opportunities influence tourism sustainability at the destination level. The analysis suggests that this system could be changed to enhance tourism sustainability by altering the nature and functions of DMOs in three major ways. Firstly, DMOs should focus on building community capacity to support a more central role for destination residents as the primary stakeholders in tourism governance. Secondly, DMOs can use stronger relationships with destination residents to develop a better understanding of DCW needs and aspirations, and use these as the objectives for tourism. Thirdly, DMOs can engage in sustainability marketing based on more sustainable tourist experience opportunities and directing promotional activities to market segments that are better matched to DCW objectives. The system outlined in Figure 6.1 can also be changed through the adoption of more contemporary planning approaches that give greater control to destination residents.

The system might also be changed by more critical research into the covert assumptions made about tourism and the social pressures on people to travel. This is especially critical as the global tourism system comes under pressure from rapid and substantial increases in new tourists from China and India. The available evidence suggests that many of these new tourists are driven by the opportunity that tourism offers to enhance social status through conspicuous consumption (Huang & Wang, 2018; Roos, 2017). This overt recognition of tourism as conspicuous consumption challenges tourism research narratives about transformative tourism, slow tourism and enlightened mass tourism.

In order to support these changes tourism research needs to be more critical and reflective, to address a number of gaps and be more innovative in its approaches.

Significant gaps include understanding the processes that link tourism to social and cultural impacts on destinations, sustainable tourism markets, and the characteristics of those who can but don't travel. In terms of innovation, tourism research needs to move away from a heavy reliance on Mode 1 forms of knowledge production and move to incorporate Mode 2 approaches to knowledge. Mode 2 knowledge production refers to research that focuses on problem solving and application of results, and is co-created with practitioners and others who live with the problem of interest (Veit et al., 2017). Finally, this chapter has focused on sustainability at the destination level and it is important to remember that systems are hierarchical and destination systems are embedded in regional and global ones. Any consideration of tourism sustainability as a whole must also recognize the importance of sustainability issues at these levels beyond the destination.

References

Abson, D. J., J. Fischer, J. Leventon, J. Newig, T. Schomerus, U. Vilsmaier, H. von Wehrden, P. Abernethy, C. D. Ives, N. W. Jager and D. J. Lang (2017), 'Leverage points for sustainability transformation', *Ambio*, **46**(1): 30–39.

Alexis, P. (2017), 'Over-tourism and anti-tourist sentiment', *Ovidius University Annals, Economic Sciences Series*, **17**(2), accessed 2 February 2018 at http://stec.univ-ovidius.ro/html/anale/ENG/2017-2/Section%20III/25.pdf.

Arnold, R. D. and J. P. Wade (2015), 'A definition of systems thinking: a systems approach', *Procedia Computer Science*, **44**: 669–78.

Bai, X., A. Surveyer, T. Elmqvist, F. W. Gatzweiler, B. Guneralp, S. Parnell, A. Prieur-Richard, P. Shrivastava, J. G. Siri, M. Stafford-Smith, J. Toussaint and R. Webb (2016), 'Defining and advancing a systems approach for sustainable cities', *Current Opinion in Environmental Sustainability*, **23**: 69–78.

Baker, M., S. Hincks and S. Sherriff (2010), 'Getting involved in plan making', *Environment and Planning C*, **28**(4): 574–94.

Bandarage, A. (2013), *Sustainability and Well-Being*. Basingstoke: Palgrave Macmillan.

Bello, F. G., B. Lovelock and N. Carr (2017), 'Constraints of community participation on protected area-based tourism planning', *Journal of Ecotourism*, **16**(2): 131–51.

Belz, F.-M. and K. Peattie (2011), *Sustainability Marketing*. Chichester: John Wiley & Sons.

Bramwell, B., J. Higham, B. Lane, and G. Miller (2017), 'Twenty-five years of sustainable tourism and the *Journal of Sustainable Tourism*: looking back and moving forward', *Journal of Sustainable Tourism*, **25**(1): 1–9.

Briggs, D. and A. Ellis (2017), 'The last night of freedom: consumerism, deviance and the "stag party"', *Deviant Behavior*, **38**(7): 756–67.

Brown, C. (2012), 'Tragedy, "tragic choices" and contemporary international political theory', in T. Erskine and R. N. Lebow (eds.), *Tragedy and International Relations*, pp. 75–85. New York: Palgrave Macmillan.

Buckley, R. (2012), 'Sustainable tourism: research and reality', *Annals of Tourism Research*, **39**(2): 528–46.

Buhalis, D. and A. Amaranggana (2014), 'Smart tourism destinations', in I. Tussyadiah and A. Inversini (eds.), *Information and Communication Technologies in Tourism 2014*, pp. 553–64. Lucerne: Springer International Publishing.

Butcher, J. (2012), 'The mantra of "community participation" in context', in T. V. Singh (ed.), *Critical Debates in Tourism*, pp. 102–08. Bristol: Channel View.

Costanza, R., J. C. Cumberland, H. E. Daly, R. Goodland and R. Norgaard (2010), *Introduction to Ecological Economics*. Boca Raton, FL: St. Lucie Press.

Crick, M. (1989), 'Representations of international tourism in the social sciences', *Annual Review of Anthropology*, **18**: 307–44.

Deery, M., L. Jago and L. Fredline (2012), 'Rethinking social impacts of tourism research', *Tourism Management*, **33**(1): 64–73.

Dwyer, L. (2018), 'Saluting while the ship sinks: the necessity for tourism paradigm change', *Journal of Sustainable Tourism*, **26**(1): 29–48.

Fisher, R. and E. Shragge (2017), 'Resourcing community organizing', *Community Development Journal*, **52**(3): 454–69.

Getz, D. (1986), 'Models in tourism planning', *Tourism Management*, **7**: 21–32.

Glackin, S. and M. R. Dionisio (2016), '"Deep engagement" and urban regeneration', *Land Use Policy*, **52**: 363–73.

Golby, J. (2018), 'Nobody actually enjoys stag and hen dos', *The Guardian*, accessed 8 July 2018 at https://www.theguardian.com/commentisfree/2015/jun/24/stag-hen-dos-york-anti-social-behaviour.

Gössling, S. and P. Peeters (2015), 'Assessing tourism's global environmental impact 1900–2050', *Journal of Sustainable Tourism*, **23**(5): 639–59.

Green, G. P. and A. Haines (2012), *Asset Building and Community Development*, 3rd edn. Los Angeles, CA: Sage.

Gunnarsdottir, N. (n.d.), '7 things Icelanders hate about tourism in Iceland', *Guide to Iceland*, accessed 8 July 2018 at https://guidetoiceland.is/history-culture/7-reasons-icelanders-hate-tourism-in-iceland.

Hall, C. M. (2010), 'Changing paradigms and global change: from sustainable to steady-state tourism', *Tourism Recreation Research*, **35**(2): 131–43.

Hanna, P., X. Font, C. Scarles, C. Weeden and C. Harrison (2018), 'Tourist destination marketing', *Journal of Destination Marketing & Management*, **9**: 36–43.

Hewlett, D. and J. Edwards (2013), 'Beyond prescription: community engagement in the planning and management of National Parks as tourist destinations', *Tourism Planning & Development*, **10**(1): 45–63.

Hodgson, A. and G. Midgley (2014), 'Bringing foresight into systems thinking', *Proceedings of the 58th Meeting of ISSS, Washington, DC, USA, July*. Accessed 12 June 2018 at http://journals.isss.org/index.php/proceedings58th/article/view/2278/770.

Howard, J. and J. Wheeler (2015), 'What community development and citizen participation should contribute to the new global framework for sustainable development', *Community Development Journal*, **50**(4), 552-570.

Huang, Z. and C. L. Wang (2018), 'Conspicuous consumption in emerging market', *Journal of Business Research*, **86**: 366–73.

Iwanicki, G., A. Dluzewska and M. S. Kay (2016), 'Assessing the level of popularity of European stag tourism destinations', *Quaestiones Geographicae*, **35**(3), accessed 8 July 2018 at https://www.degruyter.com/downloadpdf/j/quageo.2016.35.issue-3/quageo-2016-0023/quageo-2016-0023.pdf.

Jafari, J. (1990), 'The basis of tourism education', *Journal of Tourism Studies*, **1**: 33–41.

Jamrozy, U. (2007), 'Marketing of tourism: a paradigm shift toward sustainability', *International Journal of Culture, Tourism and Hospitality Research*, **1**(2): 117–30.

Jordan, E. J. and D. M. Spencer (2017), 'Perceived quality of life impacts and tourism-related stress', paper presented at the Tourism and Travel Research International Conference, Quebec City, June. Accessed 12 June 2018 at https://scholarworks.umass.edu/ttra/2017/Academic_Papers_Oral/4/.

Lenzen, M., Y. Y. Sun, F. Faturay, Y. P. Ting, A. Geschke & A. Malik (2018), 'The carbon footprint of global tourism', *Nature Climate Change*, **8**(6): 522–8.

McCool, S. F., W. A. Freimund and C. Breen (2015), 'Benefiting from complexity thinking', in G. L. Worboys, M. Lockwood, A. Kothari, S. Feary and I. Pulsford (eds.), *Protected Area Governance and Management*, pp. 291–326. Canberra: ANU Press.

Mak, B. K. L., L. T. O. Cheung and D. L. H. Hui (2017), 'Community participation in the decision-making process for sustainable tourism development in rural areas of Hong Kong, China', *Sustainability*, **9**: 1695–1708.

Marzuki, A. and I. Hay (2013), 'Towards a public participation framework in tourism planning', *Tourism Planning & Development*, **10**(4): 494–512.

Merali, Y. and P. Allen (2011), 'Complexity and systems thinking', in P. Allen, S. Maguire and B. Mckelvey (eds.), *The Sage Handbook of Complexity and Management*, pp. 31–52. London: Sage.

Mihalic, T. (2016), 'Sustainable-responsible tourism discourse – Towards "responsustable" tourism', *Journal of Cleaner Production*, **111**: 461–70.

Mingers, J. and L. White (2010), 'A review of the recent contribution of systems thinking to operational research and management science', *European Journal of Operational Research*, **207**: 1147–61.

Moscardo, G. (2008), 'Sustainable tourism innovation', *Tourism Review International*, **8**(1): 4–13.

Moscardo, G. (2009), 'Tourism and quality of life', *Tourism and Hospitality Research*, **9**(2): 159–71.

Moscardo, G. (2011), 'Exploring social representations of tourism planning', *Journal of Sustainable Tourism*, **19**(4–5), 423–36.

Moscardo, G. (2013), 'Marketing and sustainability', in G. Moscardo, G. Lamberton, G. Wells, W. Fallon, ... and W. Kershaw (eds.), *Sustainability in Australian Business: Principles and Practice*, pp. 273–314. Milton, NSW: John Wiley & Sons.

Moscardo, G. (2015a), 'Stories of people and places', in C. M. Hall, S. Gössling, and D. Scott (eds.), *The Routledge Handbook of Tourism and Sustainability*, pp. 294–301. Abingdon: Routledge.

Moscardo, G. (2015b), 'Social capital, trust and tourism development', in R. Nunkoo and S. L. J. Smith (eds.), *Trust, Tourism Development and Planning*, pp. 64–85. Abingdon: Routledge.

Moscardo, G. (2017a), 'Sustainable luxury in hotels and resorts', in M. A. Gardetti (ed.), *Sustainable Management of Luxury: Environmental Footprints and Eco-Design of Products and Processes*, pp. 163–89. Singapore: Springer.

Moscardo, G. (2017b), 'Stories as a tourist experience design tool', in D. R. Fesenmaier and X. Zheng (eds.), *Design Science in Tourism: Foundations of Destination Management*, pp. 97–124. Basel: Springer International.

Moscardo, G. (2019), 'Rethinking the role and practice of destination community involvement in tourism planning', in K. Andriotis, D. Stylidiis and A. Weidenfeld (eds.), *Tourism Policy and Planning: Issues and Challenges in Implementation*, pp. 36–52. London: Routledge.

Moscardo, G., E. Konovalov, L. Murphy, N. G. McGehee and A. Schurmann (2017), 'Linking tourism to social capital in destination communities', *Journal of Destination Marketing & Management*, **6**(4): 286–95.

Moscardo, G. and L. Murphy (2014), 'There is no such thing as sustainable tourism', *Sustainability*, **6**(5): 2538–61.

Moscardo, G. and L. Murphy (2016), 'Using destination community wellbeing to assess tourist markets', *Journal of Destination Marketing and Management*, **5**: 55–64.

Oliver, B. and B. Pitt (2013), *Engaging Communities and Service Users*. Basingstoke: Palgrave Macmillan.

Pomering, A. M., G. Noble and L. W. Johnson (2011), 'Conceptualising a contemporary marketing mix for sustainable tourism', *Journal of Sustainable Tourism*, **19**(8): 953–69.

Rasoolimanesh, S. M., M. Jaafar, A. G. Ahmad and R. Barghi (2017), 'Community participation in World Heritage Site conservation and tourism development', *Tourism Management*, **58**: 142–53.

Read, C. (2018), 'Spain cracks down on tourists: stag and hen parties threatened with £2,600 fine', *Express*, accessed 8 July 2018 at https://www.express.co.uk/news/world/959644/spain-seville-stag-hen-parties-fine.

Rodriguez-Diaz, M. and T. F. E. Rodriguez (2016), 'Determining the sustainability factors and performance of a tourism destination from the stakeholders' perspective', *Sustainability*, **8**: 951–68, doi:10.3390/su8090951.

Roos, H. (2017), 'The new economy as a gateway to leisure travelling', *Journal of Leisure Studies*, **36**(2): 192–202.

Ruhanen, R., C. Moyle and B. Moyle (2018), 'New directions in sustainable tourism research', *Tourism Review*, https://doi.org/10.1108/TR-12-2017-0196.

Sanchez-Fernandez, R., M. A. Iniesta-Bonille and A. Cervera-Taulet (2018), 'Exploring the concept of perceived sustainability at tourist destinations', *Journal of Travel & Tourism Marketing*, https://doi.org/10.1080/10548408.2018.1505579.

Saufi, A., D. O'Brien and H. Wilkins (2014), 'Inhibitors to host community participation in sustainable tourism development in developing countries', *Journal of Sustainable Tourism*, **22**(5): 801–20.

Scott, K. (2012), *Measuring Wellbeing: Towards Sustainability*. Abingdon: Routledge.

Sharpley, R. (2014), 'Host perceptions of tourism', *Tourism Management*, **42**: 37–49.

Shermer, M. (2014), 'How the survivor bias distorts reality', *Scientific American*, accessed 8 July 2018 at https://www.scientificamerican.com/article/how-the-survivor-bias-distorts-reality/.

Swapan, M. S. H. (2016), 'Who participates and who doesn't?', *Cities*, **53**: 70–77.

Thurnell-Read, T. (2011), 'Off the leash and out of control', *Sociology*, **45**(6): 977–91.

Tourism NT (2011), 'Uluru-Kata Tjuta National Park: Central Australia visitor profile and satisfaction survey', *Tourism NT*, accessed 8 July 2018 at www.tourismnt.com.au/.../uluru-kata-tjuta-national-park-visitor-profile_northern-territory_australia%20.pdf.

Tribe, J. (1997), 'The indiscipline of tourism', *Annals of Tourism Research*, **24**(3): 638–57.

Veit, D. R., D. P. Lacerda, L. F. R. Camargo, L. M. Kipper and A. Dresch (2017), 'Towards Mode 2 knowledge production', *Business Process Management Journal*, **23**(2): 293–328.

Zanini, S. (2017), 'Tourism pressures and depopulation in Cannaregio', *Journal of Cultural Heritage Management and Sustainable Development*, **7**(2): 164–78.

7 The climate change transformation of global tourism

Daniel Scott

Introduction

In its most recent scientific assessment, the United Nations (UN) Intergovernmental Panel on Climate Change (IPCC) concluded that global climate change is 'unequivocal' and that the human influence on the climate system is clear, with impacts on natural and human systems observed on all continents and the oceans (IPCC 2014).

Evidence of the rapid warming of the planet is unrelenting (USGCRP 2017). Land and sea surface temperatures have continued the multi-decade warming trend, with 14 of the 15 warmest years on record occurring in the twenty-first century (Blunden & Arndt 2017) and the global average surface temperature approximately 1.2°C above the pre-industrial era in 2016 (World Meteorological Organization 2017). Extreme temperatures have also increased over the last three decades, as human influence on the climate has increased the probability and magnitude of heat waves (Diffenbaugh, Singh & Mankin 2018; IPCC 2013). Although media headlines tend to focus on overland temperatures, they are only one indicator of our changing climate. More than 90 percent of additional energy accumulated from the human enhanced greenhouse effect is stored in the oceans. The upper layers of the oceans have warmed at a rate 50 percent faster than previously over the last four decades of the twentieth century (Durack et al. 2014).

Atmospheric concentrations of greenhouse gases (GHG) continue to increase at rates unprecedented in the geologic record. From 2015 to 2016 global carbon dioxide levels rose faster than ever before in the 58-year measurement record, surpassing the 400 parts-per-million (ppm) threshold, a level that last occurred over three million years ago (IPCC 2013). The global oceans have also absorbed much of the carbon dioxide released by human activity, increasing ocean acidity by 30 percent (Friedrich et al. 2012). The current rate of ocean geochemical changes is unprecedented over the last 300 million years, raising the possibility of far-reaching marine ecosystem change (Hönisch et al. 2012).

The global climate system will continue to respond to the elevated levels of heat-trapping GHG and ongoing historic high levels of emissions, so that additional future climate change is unavoidable. The present-day emissions rate means that there is no climate analog for this century any time in at least the last 50 million years (IPCC 2013). The magnitude of accelerating climate change and associated risks in the decades and centuries ahead will be determined by the choices to reduce GHG emissions made over the next 30 years.

Despite 20 years of international climate negotiations under the United Nations Framework Convention on Climate Change global GHG emissions rose to a historic high in 2017 (International Energy Agency 2018). The IPCC (2013) estimated that a continuation of the current high emissions trajectory would increase global mean surface temperatures 2.0°C by mid-century and 4.3°C by the end of the twenty-first century. More recent studies have increased that estimate to 4.8°C by 2100 (Brown & Caldeira 2017).

It is difficult to overstate the significance of a +4°C world. According to the IPCC (2014, p. 8) such a climate future would cause "... severe, pervasive and irreversible impacts for people and ecosystems." Little in society would remain unaffected as large areas of the earth would be transformed, greatly increasing extinction risks, exacerbating regional water and food insecurity, displacing millions of people and reducing global economic growth (IPCC 2014; New et al. 2011; World Bank 2013).

UN Secretary-General António Guterres has referred to climate change as an existential threat to humanity (UN News 2018) and climate change is one of only two global threats that has moved the Doomsday clock forward multiple times—the other being nuclear war. There are also very strong interconnections between climate change and the UN Sustainable Development Goals (SDGs) 2030 agenda. Both the IPCC (2014) and World Bank (2016) warn that climate change is already eroding the basis for sustainable development in some regions and no scenario exists by which the SDGs of 2030 could be met in a world transformed by climate change. The UN General Assembly (2017) High-Level Event on Climate Change and the Sustainable Development Agenda concluded that, "the climate and sustainable development agendas are more than mutually enforcing, their fates are intertwined." In simple terms, to fail on climate change means to fail on the SDGs.

The imperative to respond to the grand challenge of climate change is clear among global leaders of government, business and civil society. The Paris Climate Agreement, signed by 195 countries in December 2015, is the manifestation of a global consensus to set the world on a new path of international collaboration on climate change. Recognizing the consequences of even a lower magnitude of climate change the Paris Agreement (United Nations 2016, p. 22) increased global ambitions for climate stabilization to "well below 2 °C (...) and [to] pursue efforts to limit the temperature increase to 1.5 °C." The IPCC (2014) estimated global emissions will need to be reduced 40–70 percent by mid-century and that net CO_2 emissions must then decrease to zero by approximately 2070 to remain within the +2° climate guardrail. The window of opportunity to do so is very rapidly closing (Millar et al. 2017). While current pledges to the Paris Agreement are less than what is needed to achieve this policy target, the commitment to a timeline for 'peak carbon' and eventual decarbonization of the global economy has been more clearly defined and accelerated.

The implications of climate change and its impacts on environmental and socio-economic systems for the global tourism sector are far-reaching (Scott, Gössling & Hall 2012; Scott, Hall & Gössling 2016). It goes without saying that any phenomenon that will adversely affect economic growth in many areas of the world, greatly increase regional water and food insecurity, harm or displace more than a billion people, greatly increase extinction risks, increase fossil fuel-based transportation costs and progressively threaten security is not compatible with sustainable tourism development.

Regardless of whether the current high emission trajectory persists and results in a +4°C world, or a +2°C world is achieved through unprecedented deep emission reductions to a net-zero state within 50 years, both outcomes would have transformative influences on the competitiveness, sustainability and geography of tourism in the twenty-first century. Neither is a future the tourism sector understands well or is prepared for.

Tourism scholars and professionals have many questions about the implications of climate change and related mitigation and adaptation responses for the future of tourism development. Importantly, so do investors, insurers, economic development officials, community planners and many other tourism stakeholders worldwide. The World Travel and Tourism Council (WTTC) (2015, p. 5) stated that, "The next 20 years [*and beyond*] will be characterized by our sector fully integrating climate change and related issues into business strategy, supporting the global transition to a low carbon economy, strengthening resilience at a local level against climate risks . . . ." The knowledge requirements of doing so are truly massive.

This chapter will outline the pathways by which climate change and climate policy will transform aspects of global tourism. It will then identify some critical knowledge gaps and research needed to enable tourism to be part of the decarbonized and climate-resilient economy of the mid-twenty-first century and beyond.

Understanding sustainable tourism in an era of climate change

Tourism is recognized as a sector that is highly sensitive to the impacts of climate change, including environmental and socio-economic change influenced by climate change, and is also a growing contributor to anthropogenic climate change. The literature examining the complex interactions between climate change and the components of the global tourism system began in the mid-1980s, with a focus on the impacts of climate change on high-risk tourism destinations and sub-sectors. Recognition of the two-way relationship and the contribution of tourism to climate change emerged in the 1990s, although the first estimate of tourism sector greenhouse gas emissions was not published until 2008. The volume of climate change

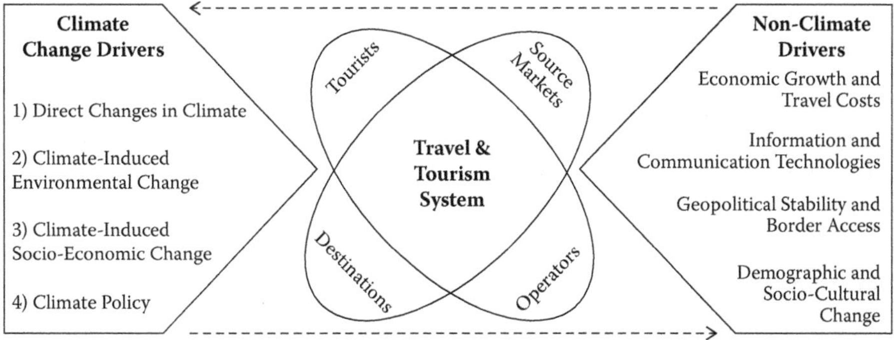

Figure 7.1 Climate change impact pathways on the tourism system

and tourism publications doubled in the 1990s and again in the 2000s (Becken 2013; Scott, Wall & McBoyle 2005), with the diversity of disciplinary contributions increasing substantially. Fang, Yin & Wu's (2017) analysis found 976 academic publications related to climate change and tourism between 1990 and 2015, with more than 60 percent published between 2010 and 2014.

The notion of a tourism system is extremely important when considering the impacts of climate change. Figure 7.1 illustrates the multi-faceted interface between climate change and different components and scales of the tourism system. Scott, Gössling & Hall (2012) identify four broad pathways through which climate change will alter the prospects for sustainable tourism development:

(1) *Direct changes in climate conditions and extreme events,* which alter the length and quality of climate-dependent tourism seasons, tourist demand, operating costs, location decisions and design, as well as damage tourism infrastructure and cause business interruptions;
(2) *Indirect climate-induced environmental changes,* which affect natural heritage assets that are critical attractions for tourists, create environmental conditions that can deter tourists, and alter operating costs and the capacity of tourism firms to do business sustainably;
(3) *Indirect climate-induced socio-economic change,* which includes decreased economic growth and discretionary wealth for tourism spending, increased political instability and security risks that deter tourists, and potentially changing consumer attitudes towards travel and associated carbon footprint;
(4) *Policy responses within tourism and in other sectors,* which includes emission reduction policies that alter transportation cost structures and destination or modal choices, as well as adaptation policies related to water rights or insurance costs and availability, which have important implications for tourism development and operating costs.

Components of the global tourism sector will be differentially affected by combinations of these four impact pathways, but because of feedbacks throughout the tourism system, all destinations will need to adapt to the risks and opportunities posed by the changing climate and climate policy.

It is beyond the scope of this chapter to comprehensively examine the multiple interconnections between tourism and climate change. The remainder of this section will outline select consequences of a changing climate and climate policy that have been critically examined in the sustainable tourism literature.

Direct impacts from a changing climate

Climate has direct and significant effects on tourism operators, destinations and tourists alike (Scott & Lemieux 2010), influencing destination competitiveness and tourist demand patterns worldwide. Despite many recent climate change analog events (e.g., heatwaves, major hurricanes, wild fires) that serve as natural experiments to understand potential future impacts of similar conditions that are anticipated to occur more frequently with climate change, observed impacts on tourism markets and responses of tourism operators, destinations and tourists are not well analyzed. A range of approaches has been used to examine how climate change may redistribute climatic resources for tourism geographically and seasonally, and the potential implications for tourism demand patterns. While studies reveal generally consistent broad patterns of influence on international tourism, contrasts also reveal some important areas for future research.

Over 20 studies have utilized a version of Mieczkowski's (1985) Tourism Climate Index (TCI), which integrates multiple climate variables relevant to tourism, to examine how the distribution of climate resources for tourism could be altered over the twenty-first century. These studies reveal largely consistent patterns of change, with climatic conditions best suited for tourism activity expanding in higher latitudes throughout the spring/summer/fall tourism seasons, improving in most mountain regions in the peak summer season, and declining in many subtropical and tropical destinations in the peak summer tourism season. Despite the TCI's wide application, it has been subject to substantial critiques (de Freitas, Scott & McBoyle 2008; Scott et al. 2016) and consequently, Scott et al. (2016) concluded that use of the TCI should be replaced by newer data-driven indices.

Studies using econometric approaches and temperature as a proxy for all climate resources for tourism similarly project a gradual shift in international tourism demand to higher latitude countries and higher elevation mountain regions (Bigano, Hamilton & Tol 2007; Hamilton, Maddison & Tol 2005). For example, even under +2°C global warming, Ciscar et al. (2014) projected tourism losses up to −11 percent for southern/Mediterranean Europe and gains in the UK and Scandinavia. The shift in regional patterns of demand is much more pronounced under warmer scenarios. Econometric studies are also not without criticism, as many rely on the aforementioned TCI and other limitations summarized (Gössling & Hall 2006; Rosselló-Nadal 2014).

With its strong reliance on specific climatic conditions, the ski industry is regarded as the tourism market most directly and immediately affected by climate change. Studies in over 20 countries have projected decreased reliability of ski slopes dependent on natural snow, increased snowmaking requirements, shortened and more variable ski seasons, a contraction in the number of operating ski areas, altered competitiveness among and within regional ski markets, increased congestion and development pressures at climate advantaged ski destinations, and attendant implications for ski tourism employment and values of vacation property real-estate values (Scott, Gössling & Hall 2012; Steiger et al. 2017). In all regional markets, the extent and timing of these impacts depends on the magnitude of climate change and the types of adaptive responses by the ski industry, skiers and destination communities. Although this is one of the most developed research areas related to climate change and tourism, a number of important limitations remain (see Steiger et al. 2017 for an overview). Modeling of future ski seasons has often utilized inappropriate impact indicators and not incorporated snowmaking, thus omitting an integral climate adaptation of the ski industry. Important limitations also exist with regard to understanding how ski tourists will respond to an evolving marketplace under climate change. Hypothetical future scenarios of snow-deficient winters have not been well defined and a large gap exists between stated and revealed behavioral responses (Rutty et al. 2015).

Climate-induced environmental change

The tourism sector is already being impacted by climate-induced changes in environmental systems that are critical assets for tourism, including biodiversity, freshwater resources, glaciers and other types of environmental heritage, including agricultural products like wine. There is also some evidence that the prevalence of observed impacts on tourism assets is contributing to the development of 'last chance' tourism markets where travelers visit destinations before they are substantially degraded by climate change impacts (e.g., polar bears, bleached coral reefs) or to witness the impacts climate change is having on landscapes (e.g., melting glaciers, wildfire burned landscapes) (Lemelin et al. 2010; Piggot-McKellar & McNamara 2017; Stewart et al. 2016).

One type of environmental change deserves particular attention because of its threat to the largest global tourism segment—coastal tourism. Sea level rise (SLR) is one of the most certain and prominent climate-induced changes to global environmental systems. Global SLR has been accelerating in recent decades and accumulating evidence indicates that under the current trajectory of high GHG emissions, global average sea-level rise could be as high as 1.5m to 2m by 2100 (Kopp et al. 2017), not the half to 1m cited in the IPCC fifth assessment report (2014). Despite the high value of tourism properties and economic activity generated by coastal tourism around the world, remarkably little analysis has been undertaken of the implications for the tourism sector.

Risks to coastal tourism infrastructure and beach assets have not been quantified in most coastal destinations. Available studies all point to transformative impacts for coastal tourism. Stanton and Ackerman (2007) estimated that with 70cm of SLR, impacts to the tourism sector in Florida would include half of existing beaches, 74 airports and over 1,300 hotels, with estimated losses in tourism properties and visitation in the tens of billions of dollars by 2050. A similar study of five major tourism beach areas in California estimated that a 1.4m SLR could cause over US$2 billion in damage by 2100 (King, McGregor & Whittett 2011). On the island of Martinique, Schleupner (2008) found that a large majority of tourist beaches (83 percent) would be affected by coastal flooding and up to 62 percent of coastal infrastructure (tourism and non-tourism) would be at risk to damage by SLR-induced erosion. A broader study of the impact of a 1m sea-level rise in 19 Caribbean countries found 29 percent of 900 coastal resorts would be partially or fully inundated, with a substantially higher proportion (49–60 percent) vulnerable to associated coastal erosion (Scott, Sim & Simpson 2012). An analysis of SLR risk to 130 UNESCO World Heritage sites found that even if global temperatures were limited to +2°C, about 19 percent (136 sites) would eventually be impacted or lost (Marzeion and Levermann 2014).

Relatively few studies have examined the potential impact of beach loss on coastal tourism demand. Surveys of tourists indicate relatively large changes in destination choice in response to degraded beaches (Uyarra et al. 2005), while interviews with tourists about their experience with ongoing erosion suggest a more diverse and tempered response (Buzinde et al. 2010). Much more is needed to assess this pre-eminent long-term threat to the sustainability of coastal tourism that is so important to economic development in Small Island Developing States and indeed coastal tourism destinations worldwide.

Climate-induced socio-economic change

Tourism depends on economic prosperity and socio-political stability, and climate change is anticipated to reduce future economic growth and contribute to political instability in some nations (IPCC 2014). Over the past decade, climate-related risks, including extreme weather events and the failure of climate change mitigation and adaptation, have ranked at the top of the World Economic Forum's Global Risk Report (2018). Burke, Hsiang and Miguel (2015, p. 235) conclude that, "If future [climate] adaptation mimics past adaptation, unmitigated warming is expected to reshape the global economy by reducing average global incomes roughly 23% by 2100 and widening global income inequality, relative to scenarios without climate change." The OECD (2014) estimated that the annual impact of only 2.5°C warming on the global economy would range between 0.7 percent and 2.5 percent of GDP to 2060 and exceed 3 percent beyond mid-century. By comparison, a 1.4 percent reduction in annual global GDP in the United Nations World Tourism Organization's (UNWTO) (2011) slower-than-expected economic growth scenario results in a large reduction (−22 percent) in international tourism arrivals in 2030 (1.4 billion instead of 1.8 billion). The extent to which a climate-disrupted global economy would impact the development trajectory of tourism remains under-researched.

Climate change influences patterns of migration and displacement and is considered a national and international security risk (Hsiang, Burke & Miguel 2013; Theisen 2017) that will intensify, particularly under greater warming scenarios. The World Bank (Rigaud et al. 2018) estimates that water scarcity, crop failure and rising sea levels may cause as many as 143 million people to be displaced from the three regions of sub-Saharan Africa, South Asia and Latin America by 2050. These same three areas are expected to be among the higher growth tourism regions, according to UNWTO (2011) projections. Displacement of large numbers of people is considered an important contributor to social instability and heightened security risks, which significantly diminish tourism activity. When awarding the Nobel Peace Prize to the IPCC in 2007, the Nobel committee recognized climate change as a global security issue, which can amplify well-documented drivers of conflict, such as poverty and economic shocks. The growing understanding of the negative social consequences of climate change needs to be translated into regional implications for tourism.

Climate mitigation and adaptation policy

Tourism and travel contribute to climate change and as the sector acts (or is regulated) to reduce its emissions consistent with the Paris Climate Agreement there will be implications for tourism costs and mobility patterns. Tourism cannot be considered sustainable unless it can be eventually decarbonized to a level consistent with a net-zero economy of the later decades of this century, and closing the emissions gap represents a major challenge for future tourism development. Indeed, it is 'peak carbon' that may restrict future tourism development, not 'peak oil' (Scott et al. 2015).

Studies commissioned by the United Nations World Tourism Organization, United Nations Environment Programme, and World Meteorological Organization (2008) and World Economic Forum (2009) estimated the contribution of the tourism sector to global anthropogenic emissions of CO_2 at approximately 5 percent in 2005. There has been considerable growth in the sector over the last decade and Lenzen et al. (2018) estimate the contribution of tourism to global CO_2 emissions at about 8 percent. Although the contribution of the tourism sector to climate change is currently considerable, the major challenge is the sector's massive projected growth and resultant future emissions trajectory. Based on a business-as-usual (BAU) growth scenario, UNWTO et al. (2008) projected CO_2 emissions from tourism would grow approximately 135 percent by 2035.

The incompatibility of this emissions growth trend with global mitigation requirements to stay within the +2°C guardrail was recognized by the IPCC (2014) Fifth Assessment Report and by the sector itself (e.g., in the 2003 Djerba and 2007 Davos Declarations on Tourism and Climate Change). In 2009, the World Travel and Tourism Council announced an aspirational emission reduction target for the sector of −50 percent by 2035 (from 2005 levels). This ambition was later endorsed by the UNWTO and is consistent with science-based emission reduction pathways recommended by the IPCC (2014) and Paris Agreement.

Although the Paris Climate Agreement placed much greater demands on GHG emission transparency and accuracy in reporting and ambition setting, the lack of systematic monitoring capabilities in tourism remains a barrier to knowing whether the sector is making progress toward its ambitions or not (Scott et al. 2015). A review of carbon reporting in travel and tourism by Bobes and Becken (2016, p. 7) found that while "(t)here is evidence that an increasing number of Travel and Tourism companies are engaging in environmental and carbon reporting ... reporting levels are still comparatively low, and quality is often insufficient."

There remains a wide gap between BAU emissions projections and the sector's stated ambitions, and no credible plan exists to close this gap (Scott, Peeters & Gössling 2010; Scott et al. 2015). Closing the gap will come at a cost. Scott et al. (2015) compared the potential costs associated with different policy pathways to achieve tourism sector emission reduction ambitions, and found investment in emissions abatement within the tourism sector combined with strategic external carbon offsets was the most cost effective over the simulation period (2015–50). The cost to achieve the −50 percent target through abatement and strategic offsetting, while significant, represented less than 0.1 percent of the estimated total global tourism economy in 2020 (approximately US$600 million annually) and grew 3.6 percent in 2050 (approximately US$350 billion annually) as emission reductions became more difficult. How this cost of decarbonization is distributed among global travelers, particularly in the aviation sector, would strongly influence its impact on destinations. Important research is needed to examine how these emission reduction ambitions might be most efficiently realized and what the potential implications for regional tourism development would be.

Notably, a review of energy-emission scenarios that achieve the Paris Climate Agreement +2°C policy target found that none explicitly considered implications for international tourism, but all assume that demand management will play a large role in achieving required emission reductions from aviation (Scott and Gössling 2015). In other words, no energy future scenario that achieves the climate change goals of the international community includes the projected growth in air travel that is central to growth scenarios of the travel and tourism sector. Tourism remains dangerously blind to the strategies that energy and climate-thought leaders, and the policy-makers they are influencing, are planning to achieve the Paris Climate Agreement low-carbon transition.

Minding the gap(s): a research agenda

In addition to the limitations and gaps within the extant literature identified in the previous section, there are a number of overarching gaps that need to be prioritized in order to advance climate change adaptation and mitigation decision making in the sector.

Regional knowledge gaps

Substantial regional imbalances and important geographic gaps exist in the available literature on climate change and tourism. The place of tourism in regional chapters of the periodic IPCC assessments remains an important indicator of progress on regional knowledge gaps (Scott, Hall & Gössling 2016). Europe and Australia/New Zealand represent the most well developed tourism discussions, summarizing evidence on a range of impacts, the level of awareness and concern of tourism stakeholders, and the potential limits of adaptation. Tourism content in the North America chapter of the Fifth Assessment declined relative to the Fourth Assessment; however, this was not due to limited research available, but rather the composition of the authorship team and the level of importance ascribed to the sector.

Africa, Asia and Central/South America have been identified as regional knowledge gaps for over a decade (Amelung, Moreno & Scott 2008; Hall 2008; Scott, Hall & Gössling 2016), and progress remains limited. Although research specific to the Asia and Africa regions is beginning to emerge, the Central/South America region is almost devoid of empirical studies of specific tourism vulnerabilities within the region. IPCC chapters for all three regions still largely identify generic impacts, without discussion of their potential magnitude or timing, or tourism sector interactions with other major impacts in the region. As Boko et al. (2007, p. 459) indicate, "... there is a need to enhance practical research regarding the vulnerability and impacts of climate change on tourism, as tourism is one of the most important and highly promising economic activities in Africa." The same can be said for Asia and the Middle East, regions that are expected to lead global growth in tourism through the 2030s (UNWTO 2011). (See Scott, Hall & Gössling's 2016 assessment of the knowledge gaps in the regional chapters of the IPCC Fifth Assessment Report for additional details.)

Transnational climate change impacts

Climate change impacts traverse political boundaries, and government or business adaptation and mitigation responses in one country can have important ramifications in many other regions. Benzie, Hedlund & Carlsen (2016) refer to this as 'transnational' risks that reach across national borders (those with common borders as well as more distant countries). Because tourism is characterized by strong global interconnectedness, the consequences of climate change and associated mitigation/adaptation responses in other countries can have important implications for destination countries (Hamilton, Maddison & Tol 2005; Scott, Gössling & Hall 2012; UNWTO et al. 2008). For example, if an international climate change mitigation policy regime increases the cost of air travel substantially enough to alter destination choice, arrivals to some destinations could be affected regardless of local-scale climate adaptation and sustainable tourism actions. Similarly, if the insurance industry deems coastal risks of SLR in an area too high to continue property insurance, the loss of insurability could negate destination-scale adaptations and sustainable coastal management practices. The choice of climate change adaptation strategies at the destination scale can potentially be heavily influenced by external factors that are not adequately represented in conventional vulnerability assessment approaches. Reisinger et al. (2014, p. 1376) indicated that, "effects from climate change impacts and responses outside Australasia have the potential to outweigh some of the direct impacts within the region, particularly economic impacts on trade-intensive sectors such as agriculture and tourism, but they remain among the least-explored issues." The need for innovative vulnerability assessment approaches capable of accounting for important transnational risks associated with international tourism, are a clear research priority.

Tourist responses to climate change

Within the tourism system, tourists have the largest capacity to adapt to the consequences of climate change and climate policy, with tremendous flexibility to substitute the place, timing and type of holiday, even at very short notice. As a consequence, understanding the potential response of tourists to changing climate and environment assets, mobility costs and socio-economic impacts is a priority to improve projections of changes in tourism demand patterns (Gössling, Scott & Hall 2012). Natural experiments of tourist responses to ongoing and future impacts of climate change that are offered by climate anomalies (e.g., record warm and snow-deficient winters, summer heat waves) and policy changes (e.g., the substantial increase of the UK Air Passenger Duty as a proxy for a carbon tax) continue to be opportunities lost. There continues to be a strategic need to address uncertainties related to tourist perceptions and demand responses to the impacts of climate change which act as an important barrier to government and business mitigation and adaptation responses at the destination level.

Cumulative and compounding climate change impacts

An important limitation of the climate change and tourism literature is that studies have tended to examine potential impacts from only one pathway and often for only one destination or market segment (Hamilton, Maddison & Tol 2005; Rosselló-Nadal 2014; Scott, Gössling & Hall 2012). If a large number of competing tourism operators or destinations are impacted, these changes and associated investor and tourist responses will alter competitiveness throughout parts of the tourism system. A negative impact in one part of the tourism system may constitute an opportunity elsewhere in the marketplace. Studies that examine possible compounding impacts from the four pathways and interactions between them and other macro-scale drivers of tourism development outlined in Figure 7.1 remain limited (Scott, Hall & Gössling 2016).

One effort to examine the collective influence of the four impact pathways and better understand the geography of climate change risk for the global tourism sector combined 27 indicators of internal and transnational climate change impacts (across all four pathways) and tourism sector and destination country adaptive capacity into a climate change vulnerability index for 181 countries (Scott and Gössling 2018). Countries with the lowest risk and potential opportunities were found in western and northern Europe, central Asia, and Canada and New Zealand. The highest sector risk was found in Africa, the Middle East, South Asia and Small Island Developing States, where tourism represents a significant proportion of the national economy and contributor to the SDGs. Climate change risk also aligned strongly with regions where tourism growth is projected to be the strongest over the coming decades, including the sub-regions of sub-Saharan Africa and South Asia. This first more holistic picture of the geography of climate change risk for tourism has important implications for the many developing countries that look to tourism as a central future development strategy, and must be more thoroughly considered in national development plans, official development assistance programs and international adaptation financing negotiations.

While this study is valuable for improving an understanding of the cumulative stress climate change will place on the tourism sector of individual countries and regions, this index approach is limited in that it is summative and does not account for the interactions between the suite of impacts that may exacerbate or accelerate climate risks and opportunities. New research is required to develop techniques that analyze interconnections among climate impacts, particularly at the destination scale, and incorporate this information into tourism planning.

Toward climate change ready tourism

Both the UNWTO and WTTC acknowledge that climate change is already impacting the tourism sector. As researchers we observe the accumulating evidence, evaluate media accounts, and gain insights through conversations with professionals throughout the sector, as well as other tourism stakeholders like investors or policy-makers that seek expert advice on managing impacts and proactive responses. What we are witnessing is but the very early stages of what is to come, as a myriad of unknown future challenges will arise from accelerating climate change in the coming decades. Growing evidence indicates that the magnitude of projected impacts are dependent on the scale of changes in climate and that sector risks will be much greater under higher emission and higher temperature increases (Jones and Phillips 2017; Scott, Gössling & Hall 2012; Steiger et al. 2017).

All tourism destinations will need to adapt to climate change, whether to manage risks or to capitalize on new opportunities associated with local impacts of climate change or impacts on competitors and the global tourism system. Unfortunately, the preparedness of the tourism sector for the current and future challenges of climate change has repeatedly been called into question (KPMG 2008; OECD and UNEP 2011; Scott and Gössling 2018; Scott, Wall & McBoyle 2005). KPMG's (2008) assessment of the regulatory, physical, reputational and litigation risks of climate change posed to 18 major economic sectors versus their level of preparedness, found tourism to be one of six sectors in the 'danger zone' (along with aviation, transport, health care, the financial sector, and oil and gas). An assessment of 18 leading tourism countries revealed that none had completed a strategic review of the vulnerabilities and potential adaptation strategies within the tourism sector (OECD and UNEP 2011). The tourism sector is mentioned in only 82 countries' Nationally Determined Contributions submitted to the Paris Climate Agreement (Scott and Gössling 2018), and is not explicitly identified by some of the countries where tourism is most at risk to climate change, signifying a need to raise awareness about sectoral climate change risks and the essential connections to advancing SDGs. Consequently, tourism remains ill-prepared to rapidly evolving demands of institutional investors and the financial sector for disclosure on carbon and physical climate risks (European Bank for Reconstruction and Development 2018; Task Force on Climate-Related Financial Disclosures 2017).

Tourism growth projections, increasing to 1.8 billion international tourist arrivals in 2030 (UNWTO 2011) and between 2.4 and 3.1 billion in 2050 (UNEP 2012), are predicated on assumptions of relative stability in global economic, social, political and environmental systems. Tourism leaders have not yet begun to envision the global tourism system in a climate-disrupted +4°C world or what international tourism could look like in a post-carbon, +2°C world. The trajectory for one of these outcomes will be set in the next 10 to 20 years.

The academy has a crucial role to inform and engage the broader tourism community in overcoming this stability bias and asking challenging questions about the long-range future of tourism development. How will the competitive positions of destinations and major tourism resort companies be re-shaped by declining snow and beach resources over time? How would the geography of global tourism change if sustainable biofuels fail to emerge in the 2020s and 2030s and the price of carbon offsets increase substantially, or anticipated unrestricted growth in air travel is curtailed by an increasingly stringent regulatory regime? What would be the implications for many countries that consider tourism as a central pillar of future economic development and the broader promise of contributing to the United Nations Sustainable Development Goals? These and many other salient questions remain largely unasked, yet are fundamental to the future of sustainable tourism.

Some scholars (e.g., Weaver 2011, p. 5) have questioned whether the field is becoming "dominated – at least rhetorically – by the issue of climate change" and if "tourism's expanding engagement with climate change, as it is currently unfolding, is not necessarily conducive to the interests of tourism sustainability." Can tourism really afford not to increase research and engagement if the sector is already considered one of the least prepared? An inadequate and late climate change response by the tourism sector will severely degrade progress on sustainable tourism and its capacity to contribute to the SDGs in the decades ahead. That climate change raises disconcerting questions about sustainable tourism is not a justification for retrenchment, but rather demands greater reflection on what forms of tourism development are compatible with a decarbonized and at least partially climate-disrupted global economy.

References

Amelung, B., Moreno, A. and Scott, D. (2008). 'The place of tourism in the IPCC Fourth Assessment Report: a review', *Tourism Review International*, **12**(1), 5–12.

Becken, S. (2013). 'A review of tourism and climate change as an evolving knowledge domain', *Tourism Management Perspectives*, **6**: 53–62.

Benzie, M., Hedlund, J. and Carlsen, H. (2016). *Introducing the Transnational Climate Impacts Index: Indicators of Country-Level Exposure Methodology Report*. Stockholm: Stockholm Environment Institute.

Bigano, A., Hamilton, J. and Tol, R. (2007) 'The impact of climate change on domestic and international tourism: a simulation study', *Integrated Assessment Journal*, **7**: 25–49.

Blunden, J. and Arndt, D. S. (eds.) (2017). 'State of the climate in 2016', *Bulletin of the American Meteorological Society*, **98**(8): Si–S277.

Bobes, L. and Becken, S. (2016). *Proving the Case: Carbon Reporting in Travel and Tourism*. Available at http://www.amadeus.com/documents/reports/carbon-reporting-taveland-tourism.pdf.

Boko, M., Niang, I., Nyong, A., Vogel, C., Githeko, A., Medany, M., Osman-Elasha, B., Tabo, R. and Yanda, P. (2007). 'Africa', in M. L. Parry, O. F. Canziani, J. P. Palutikof, P. J. van der Linden and C. E. Hanson (eds.), *Climate Change 2007: Impacts, Adaptation and Vulnerability. Contribution of Working Group II to the Fourth Assessment Report of the Intergovernmental Panel on Climate Change*, pp. 433–67. Cambridge: Cambridge University Press.

Brown, P. and Caldeira, K. (2017). 'Greater future global warming inferred from Earth's recent energy budget', *Nature*, **552**: 45–50.

Burke, M., Hsiang, S. and Miguel, E. (2015). 'Global non-linear effect of temperature on economic production', *Nature*, **527**: 235–9.

Buzinde, C., Manuel-Navarrete, D., Yoo, E. and Morais, D. (2010). 'Tourists' perceptions in a climate of change: eroding destinations', *Annals of Tourism Research*, **37**: 333–54.

Ciscar, J.-C., Feyen, L., Soria, A., Lavalle, C., Raes, F., Perry, M., et al. (2014). *Climate Impacts in Europe—The JRC 48 PESETA II Project*, ed. J. C. Ciscar. Publications Office of the European Union Available at http://publications.jrc.ec.europa.eu/repository/handle/JRC87011.

de Freitas, C., Scott, D. and McBoyle, G. (2008). 'A second-generation climate index for tourism: specification and verification', *International Journal of Biometeorology*, **52**: 399–407.

Diffenbaugh, N., Singh, D. and Mankin, J. (2018). 'Unprecedented climate events: historical changes, aspirational targets, and national commitments', *Science Advances*, **4**(2): doi: 10.1126/sciadv.aao3354.

Durack, P. J., Gleckler, P. J., Landerer, F. W. and Taylor, K. E. (2014). 'Quantifying underestimates of long-term upper-ocean warming', *Nature Climate Change*, **4**: 999–1005.

European Bank for Reconstruction and Development (2018). *Advancing Task Force on Climate-Related Financial Disclosures Guidance on Physical Climate Risks and Opportunities*. Available at https://climatecentre.org/downloads/files/EBRD-GCECA%20report.compressed.pdf.

Fang, Y., Yin, J. and Wu, B. (2017). 'Climate change and tourism: a scientometric analysis using CiteSpace', *Journal of Sustainable Tourism*, **26**(1): 108–26.

Friedrich, T., Timmermann, A., Abe-Ouchi, A., Bates, N., Chikamoto, M., Church, M., and Santana-Casiano, J. (2012). 'Detecting regional anthropogenic trends in ocean acidification against natural variability', *Nature Climate Change*, **2**(3): 167–71.

Gössling, S. and Hall, C. M. (2006). 'Uncertainties in predicting tourist travel flows based on models', *Climatic Change*, **79**:163–73.

Gössling, S., Scott, D. and Hall, C. M. (2012). 'Consumer behaviour and demand response of tourists to climate change', *Annals of Tourism Research*, **39**(1): 36–58.

Hall, C. M. (2008). 'Tourism and climate change: knowledge gaps and issues', *Tourism Recreation Research*, **33**: 339–50.

Hamilton, J., Maddison, D. and Tol, R. (2005). 'Climate change and international tourism: a simulation study', *Global Environmental Change*, **15**: 253–66.

Hönisch, B., Ridgwell, A., Schmidt, D., Thomas, E., Gibbs, S., Sluijs, A. and Williams, B. (2012). 'The geological record of ocean acidification', *Science*, **335**(6072): 1058–63.

Hsiang, S., Burke, M. and Miguel, E. (2013). 'Quantifying the influence of climate on human conflict', *Science*, **341**: 1–14.

International Energy Agency (2018). 'Global energy demand grew by 2.1% in 2017, and carbon emissions rose for the first time since 2014'. Available at https://www.iea.org/newsroom/news/2018/march/global-energy-demand-grew-by-21-in-2017-and-carbon-emissions-rose-for-the-firs.html.

IPCC (Intergovernmental Panel on Climate Change) (2013). 'Summary for policymakers'. *Climate Change 2013: The Physical Science Basis. Contribution of Working Group I to the Fifth Assessment Report of the Intergovernmental Panel on Climate Change*. Cambridge: Cambridge University Press.

IPCC (Intergovernmental Panel on Climate Change) (2014). *Climate Change 2014: Synthesis Report. Contribution of Working Groups 1, 2, and 3 to the Fifth Assessment Report of the Intergovernmental Panel on Climate Change*, ed. R. K. Pachauri and L.A. Meyer. Geneva: IPCC.

Jones, A. and Phillips, M. (eds.) (2017). *Climate Change and Coastal Tourism: Recognizing Problems, Meeting Expectations & Managing Solutions*. Wallingford: CABI.

King, P., McGregor, A. and Whittett, J. (2011). *The Economic Costs of Sea-Level Rise to California Beach Communities*. Fresno: California Department of Boating and Waterways, San Francisco State University. Available at http://www.dbw.ca.gov/PDF/Reports/CalifSeaLevelRise.pdf.

Kopp, E., DeConto, R., Bader, D., Hay, C., Horton, R., Kulp, S., Oppenheimer, M., Pollard, D. and Strauss, B. (2017). 'Evolving understanding of Antarctic ice-sheet physics and ambiguity in probabilistic sea-level projections', *Earth's Future*. Available at https://doi.org/10.1002/2017EF000663.

KPMG (2008). *Climate Changes your Business*. Amsterdam: KPMG. Available at http://www.climatebiz.com/sites/default/files/document/Climatechang_riskreport.pdf.

Lenzen, M., Sun, Y., Faturay, F., Ting, Y., Geschke, A. and Malik, A. (2018). 'The carbon footprint of global tourism', *Nature Climate Change*, **8**: 522–8.

Lemelin, H., Dawson, J., Stewart, E. J., Maher, P. and Lueck, M. (2010). 'Last-chance tourism: the boom, doom, and gloom of visiting vanishing destinations', *Current Issues in Tourism*, **13**(5): 477–93.

Marzeion, B. and Levermann, A. (2014). 'Loss of cultural world heritage and currently inhabited places to sea-level rise', *Environmental Research Letters*, **9**(3): 34001.

Mieczkowski, Z. (1985). 'The tourism climate index: a method for evaluating world climates for tourism', *The Canadian Geographer*, **29**: 220–33.

Millar, R., Fuglestvedt, J., Friedlingstein, P., Rogelj, J., Grubb, M., Matthews, H. D., Skeie, R., Forster, P., Frame, D. and Allen, M. (2017). 'Emission budgets and pathways consistent with limiting warming to 1.5 °C', *Nature Geoscience*, **10**: 741–7.

New, M., Liverman, D., Schroder, H. and Anderson, K. (2011). 'Four degrees and beyond: the potential for a global temperature increase of four degrees and its implications', *Philosophical Transactions of the Royal Society A: Mathematical, Physical and Engineering Sciences*, **369**(1934): 6–19.

OECD (2014). 'A call for zero net emissions'. Available at http://oecdinsights.org/2014/01/24/a-call-for-zero-emissions/ (accessed 1 March 2015).

OECD and UNEP (Organisation for Economic Co-operation and Development and United Nations Environment Program) (2011). *Sustainable Tourism Development and Climate Change: Issues and Policies*. Paris: OECD.

Piggott-McKellar, A. E. and McNamara, K. E. (2017). 'Last-chance tourism and the Great Barrier Reef', *Journal of Sustainable Tourism*, **25**: 397–415.

Reisinger, A., Kitching, R., Chiew, F., Hughes, L., Newton, P., Schuster, S., . . . and Whetton, P. (2014). 'Australasia', in C. B. Field, V. Barros, D. Dokken, K. Mach, M. Mastrandrea, T. Bilir, . . . and L. White (eds.), *Climate Change 2014: Impacts, Adaptation, and Vulnerability. Part B: Regional Aspects*, pp. 1371–438. Cambridge: Cambridge University Press.

Rigaud, K. K., de Sherbinin, A., Jones, B., Bergmann, J., Clement, V., Ober, K., Schewe, J., Adamo, S., McCusker, B., Heuser, S. and Midgley, A. (2018). *Groundswell: Preparing for Internal Climate Migration*. Washington, DC: World Bank. Available at https://openknowledge.worldbank.org/handle/10986/29461.

Rosselló-Nadal, J. (2014). 'How to evaluate the effects of climate change on tourism', *Tourism Management*, **42**: 334–40.

Rutty, M., Scott, D., Steiger, R., Pons, M. and Johnson, P. (2015). 'Behavioural adaptation of skiers to climatic variability and change in Ontario, Canada', *Journal of Outdoor Recreation and Tourism*, **11**: 13–21.

Schleupner, C. (2008). 'Evaluation of coastal squeeze and its consequences for the Caribbean Island Martinique', *Ocean and Coastal Management*, **51**: 383–90.

Scott, D. and Gössling, S. (2015). 'What could the next 40 years hold for global tourism?', *Tourism Recreation Review*, **40**(3): 269–85.

Scott D. and Gössling, S. (2018). *Tourism and Climate Change: Embracing the Paris Agreement*. Dublin: European Travel Commission. Available at https://etc-corporate.org/reports/tourism-and-climate-change-mitigation-embracing-the-paris-agreement/.

Scott, D., Gössling, S. and Hall, C. M. (2012). 'International tourism and climate change', *Wiley Interdisciplinary Reviews – Climate Change*, **3**(3): 213–32.

Scott, D., Gössling, S., Hall, C. M. and Peeters, P. (2015). 'Can tourism be part of the decarbonized global economy? The policy costs and risks of carbon reduction strategies', *Journal of Sustainable Tourism*, **24**(1): 52–72.

Scott, D., Hall, C. M. and Gössling, S. (2016). 'A review of the IPCC Fifth Assessment and implications for tourism sector climate resilience and decarbonization', *Journal of Sustainable Tourism*, **24**: 8–30.

Scott, D. and Lemieux, C. (2010). 'Weather and climate information for tourism', *Proceedia Environmental Sciences*, **1**: 146–83.

Scott, D., Peeters, P. and Gössling, S. (2010). 'Can tourism deliver its "aspirational" emission reduction targets?', *Journal of Sustainable Tourism*, **18**: 393–408.

Scott, D., Rutty, M., Amelung, B. and Tang, M. (2016). 'A comparison of the Holiday Climate Index and Tourism Climate Index in European urban destinations', *Atmosphere*, **7**(6): 80.

Scott, D., Sim, R. and Simpson, M. (2012). 'Sea-level rise impacts on coastal resorts in the Caribbean', *Journal of Sustainable Tourism*, **20**(6): 883–98.

Scott, D., Wall, G. and McBoyle, G. (2005). 'The evolution of the climate change issue in the tourism sector', in M. Hall and J. Higham (eds.), *Tourism, Recreation and Climate Change*, pp. 44–60. London: Channelview Press.

Stanton, E. A. and Ackerman, F. (2007). *Florida and Climate Change: The Costs of Inaction*. Medford, MA: Global Development and Environment Institute, Tufts University and Stockholm Environment Institute–US Center, Tufts University.

Steiger, R., Scott, D., Abegg, B., Pons, M. and Aall, C. (2017). 'A critical review of climate change risk for ski tourism', *Current Issues in Tourism*, doi: 10.1080/13683500.2017.1410110.

Stewart, E. J., Wilson, J., Espiner, S., Purdie, H., Lemieux, C. and Dawson, J. (2016). 'Implications of climate change for glacier tourism', *Tourism Geographies*, **18**: 377–98.

Task Force on Climate-Related Financial Disclosures (2017). *Final Report: Recommendations of the Task Force on Climate-related Financial Disclosures*. Available at https://www.fsb-tcfd.org/publications/.

Theisen, O. M. (2017). 'Climate change and violence: insights from political science', *Current Climate Change Reports*, doi: 10.1007/s40641-017-0079-5.

UN General Assembly (2017). *High Level SDG Action Event – Climate Change and the Sustainable*

Development Agenda. Available at https://www.un.org/pga/71/wp-content/uploads/sites/40/2015/08/President-Summary_SDG-Action-Event-on-Climate-and-SD_23-March-2017_FINAL.pdf.

UN News (2018). 'Climate change: an "existential threat" to humanity, UN chief warns global summit'. Available at https://news.un.org/en/story/2018/05/1009782.

UNEP (United Nations Environment Programme) (2012). *Tourism in the Green Economy – Background Report*. Available at http://www.unep.org/greeneconomy/Portals/88/documents/ger/ger_final_dec_2011/Tourism%20in%20the%20green_economy%20unwto_unep.pdf.

United Nations (2016). *United Nations Treaty Collection Chapter XXVII Environment, 7.d Paris Agreement*, Paris, 12 December 2015 (C.N.92.201). New York: United Nations. Available at https://treaties.un.org/doc/Publication/CN/2016/CN.92.2016-Eng.pdf (accessed 1 April 2016).

United Nations World Tourism Organization, United Nations Environment Programme and World Meteorological Organization (2008). *Climate Change and Tourism: Responding to Global Challenges*. Madrid & Paris: UNWO, UNEP.

UNWTO (United Nations World Tourism Organization) (2011). 'Tourism Towards 2030: Global Overview'. UNWTO General Assembly, 19th session.

USGCRP (2017). *Climate Science Special Report: A Sustained Assessment Activity of the U.S.*, ed. D. J. Wuebbles, D. W. Fahey, K. A. Hibbard, D. J. Dokken, B. C. Stewart and T. K. Maycock. Washington, DC: U.S. Global Change Research Program.

Uyarra, M., Côté, I., Gill, J., Tinch, R., Viner D. and Watkinson, A. (2005). 'Island-specific preferences of tourists for environmental features: implications of climate change for tourism-dependent states', *Environmental Conservation*, **32**(1): 11–19.

Weaver, D. (2011). 'Can sustainable tourism survive climate change?', *Journal of Sustainable Tourism*, **19**(1): 5–15.

World Bank (2013). *Turn Down the Heat: Climate Extremes, Regional Impacts, and the Case for Resilience*. Washington, DC: World Bank.

World Bank (2016). 'Climate action does not require economic sacrifice'. Available at http://blogs.worldbank.org/climatechange/climate-action-does-not-require-economic-sacrifice (accessed 30 March 2016).

World Economic Forum (2009). *Towards a Low Carbon Travel and Tourism Sector*. Davos: WEF.

World Economic Forum (2018). *The Global Risks Report 2018* (13th edn.). Davos: World Economic Forum. Available at https://www.weforum.org/reports/the-global-risks-report-2018 (accessed June 2018).

World Meteorological Organization (2017). 'WMO confirms 2017 among the three warmest years on record'. Available at https://public.wmo.int/en/media/press-release/wmo-confirms-2017-among-three-warmest-years-record.

World Travel and Tourism Council (2009). *Leading the Challenge on Climate Change*. London: WTTC.

World Travel and Tourism Council (2015). *Travel and Tourism 2015: Connecting Global Climate Action*. London: WTTC.

8 The contribution of tourism to achieving the United Nations Sustainable Development Goals

Anna Spenceley and Andrew Rylance

Overview

In 2015 governments adopted the United Nations 2030 Agenda for Sustainable Development and the Sustainable Development Goals (SDGs). The agenda established a global framework to end extreme poverty, fight inequality and injustice, and remedy climate change. Building on the Millennium Development Goals, 17 SDGs and 169 associated targets were agreed. Tourism is included within the targets for Goal 8 on decent work and economic growth; Goal 12 on responsible consumption and production, and Goal 14 on life below water. However, tourism has the potential to contribute, directly or indirectly, to all of the goals (World Tourism Organization, 2015).

The United Nations 70th General Assembly designated 2017 as the International Year of Sustainable Tourism for Development. This provided an opportunity to raise awareness of the contribution of sustainable tourism to development, while mobilizing all stakeholders to collaborate in using tourism as a catalyst for positive change. The International Year focused on five key areas, namely sustainable economic growth; social inclusiveness, employment and poverty reduction; resource efficiency, environmental protection and climate change; cultural values, diversity and heritage; and mutual understanding, peace and security (UNWTO & UNDP, 2017). A review was undertaken of 64 countries' Voluntary National Reviews of the SDGs, and corporate social responsibility activities of 60 tourism companies (UNWTO, 2017a). The findings included that SDGs 8, 12 and 17 have the strongest links with tourism, but that there are very few links between SDGs 3, 4, 7 and 10 with the sector. Furthermore, challenges relate to irresponsible consumption and production, and poor management of resources related to SDGs 12, 11 and 14. Key findings included that policy makers should encourage and support the tourism private sector, and that active engagement and coherent dialogue are required to optimize progress. For the private sector, internalization of the SDGs relates to their drive towards competitiveness and profitability, rather than philanthropy. Therefore, more inclusive and sustainable business models need to relate to core business activities.

The following section describes major dimensions of past research, coupled with examples to illustrate how tourism contributes to SDGs.

Major dimensions of past research

Goal 1. No poverty: End poverty in all its forms everywhere

In 2013, it was estimated that 10.7 per cent of the world's population lived on less than US$1.90 per day (World Bank, 2016a). However, poverty is a multidimensional phenomenon, and it manifests where people have inadequate income, a lack of access to education, poor health, insecurity, low self-confidence, a sense of powerlessness, and where there is an absence of rights (Sen, 1999).

Bennett, Ashley & Roe (1999) suggested that the tourism sector had the potential to contribute to poverty reduction in developing countries, because the market comes to the producers, inter-sectoral linkages can be created, it is labour intensive (particularly for women, youth and people will low skills), can take place in marginal areas, and it has fewer barriers to entry than manufacturing or other export activities. There has been extensive research on 'pro-poor tourism' (e.g. Ashley & Mitchell, 2007; Ashley, Roe & Goodwin, 2001; Mitchell & Ashley, 2009; Spenceley & Meyer, 2016) addressing the opportunity to harness markets for poverty reduction, and tools for doing so.

Tourism generated an estimated US$1.5 trillion in export earnings in 2015 (UNWTO, 2017b). Scheyvens (2011) estimated that approximately 40 per cent of all international tourist arrivals accrue to developing countries, and so tourism can be a significant foreign exchange earner. Some of the poorest regions of the world are rich cultural and natural assets, which offer great potential as visitor attractions. Tourism can provide a mechanism to re-distribute wealth from the rich to the poor, because as tourists travel they spend money on travel, accommodation, excursions, food, drinks and shopping (Spenceley & Meyer, 2016). For example, Pafuri Camp, a luxury lodge in South Africa, employs around 52 permanent staff members, and 94 per cent of them are from the local Makuleke community. Employees from the community collectively receive approximately US$298,000 in wages and related benefits annually, which makes a substantial impact in the local economy, and contributes to poverty reduction (Snyman & Spenceley, 2019).

Goal 2. Zero hunger: End hunger, achieve food security and improved nutrition and promote sustainable agriculture

Tourism can catalyse sustainable agriculture by promoting the production of foodstuffs and supplies to hotels and restaurants, and also through sales of local produce to tourists. Agro-tourism can generate additional income for farmers while providing rich and educational tourism experiences (UNWTO & UNDP, 2017). Agriculture and the harvesting of natural resources continues to remain a predominant livelihood opportunity for poor communities working in rural areas, accounting for 55 per cent of employment in developing countries and is the main source of income for the rural poor (Schiere & Kater, 2001). A review of 49 tropical protected areas (PAs) showed that they are becoming isolated as deforestation takes place around their boundaries and therefore effective management needs to address the wider local socio-economic developmental issues (Naughton-Treves, Holland & Brandon, 2005). Diversification strategies are important for poor communities to reduce their dependence on and the associated risks with a single income stream, such as farming (Ashley, Mdoe & Reynolds, 2002).

Marine protected areas used by tourists and fishermen provide different opportunities to reduce hunger. For example, small community-managed marine protected zones can provide (a) refuge for breeding and nursing populations of fish to support the local subsistence fishing industry, and (b) areas for non-consumptive marine tourism (e.g. whale shark viewing and manta ray diving), which provide job opportunities for local people. Such a system is being developed by a luxury lodge company, &Beyond, in collaboration with local communities and authorities in Tanzania and Mozambique (Braack & Mearns, 2017). Here, efforts are being made to tackle overfishing and the killing of endangered marine species, protect reefs and endangered species, and also build capacity among local communities through the management of local fishing stocks and responsible community fishing practices (Braack & Mearns, 2017).

Goal 3. Good health and well-being: Ensure healthy lives and promote well-being for all of all ages

There is increasing scientific evidence of the health benefits of protected areas, and 'Healthy Parks Healthy People' was one of the core themes of the International Union for Conservation of Nature (IUCN) 2014 World Parks Congress in Sydney Australia (Spenceley, 2017). Visitation to areas of high biodiversity can be a tool in preventative medicine, and provide health benefits caused by certain lifestyle problems, such as obesity, cardiovascular disease, depression and anxiety (Sparkes & Woods, 2009). In Australia, Parks Victoria formed a partnership with two major players in Australia's health care delivery system, Medibank Australia and the National Heart Foundation. This includes providing health care professionals with options to prescribe physical activity in protected areas as a proactive disease prevention approach (HPHP, 2017).

Promoting sport tourism is an increasing area of interest for PAs as both a means of generating revenue to finance conservation efforts as well as demonstrating the wider social contribution of biodiversity to local communities. For example, the vision of the Queensland Department of National Parks, Australia, is "Flourishing parks and forests. Physically active Queenslanders. Public confidence in the racing industry" (Department of National Parks, 2014). Furthermore, running events, bicycle racing, triathlons and adventure walks are increasing being utilized as means to increase visitation as well as provide wider health benefits. Some excellent examples include cross-border tourism adventure products in southern African transfrontier conservation areas, including the Tour de Tuli (see Goal 4), Desert Knights (a cross-border canoe and mountain-biking event), and Wildruns (cross-border trail-runs) (Spenceley, 2018a).

Corporate social responsibility initiatives by nature-based tourism operators working in areas of high biodiversity, and philanthropic efforts, also often contribute towards public health improvements. For example, the Africa Foundation is an organization that channels donations from andBeyond's guests and donors into social initiatives in communities neighbouring their lodges, such as Phinda Game Reserve in South Africa. Over more than two decades, donations have supported health initiatives in communities neighbouring Phinda such as construction of a clinic at Mduku, and refurbishment of another, a series of HIV/AIDS awareness workshops, improved water provision, and a school permaculture project (Snyman & Spenceley, 2019).

Goal 4. Quality education: Ensure inclusive and equitable quality education and promote lifelong learning opportunities for all

A skilled workforce is crucial in order to provide quality hospitality and experiences to tourists. The tourism sector provides professional development and training opportunities for direct and indirect jobs for youth, women, and those with special needs (UNWTO & UNDP, 2017). For example, one of !Xaus Lodge's cleaning staff in South Africa joined the lodge soon after it opened and developed a special interest in guiding. Encouraged by the lodge managers, Melissa Mienies taught herself key information (despite having failed her secondary school exams). !Xaus Lodge also supported her with a distance learning course, and she qualified as a guide. She was the first female nature guide in the Kgalagadi who is a member of the local community (Snyman & Spenceley, 2019).

Tourism also provides opportunities for environmental education of tourists, and of local community members and youth. For example Wilderness Safaris has a non-profit organization called Children in the Wilderness (CITW), which since 2005 has organized an annual international cycling event within the Mapungubwe Transfrontier Conservation Area called the Tour de Tuli. Between 2005 and 2017, donations from 3,770 cyclists participating in the tour raised over US$1.7 million. At the end of 2016, these funds had been used for over 5,600 children to attend a CITW environmental awareness camp at a Wilderness Safaris lodge, and for 6,000 children to participate in an Eco-Club programme (Spenceley, 2017).

Goal 5. Gender equality: Achieve gender equality and empower all women and girls

Research from the World Bank has found that women lag behind men in nearly all measures of economic opportunity in the world (World Bank, 2016b) and that the inequalities are most stark in low-income countries (Twining-Ward & Zhou, 2017). However, tourism can empower women, particularly through the provision of direct jobs and income-generation from small, medium and micro enterprises (SMMEs) in tourism and hospitality-related enterprises (UNWTO & UNDP, 2017). Characteristics of the tourism sector that may explain the stronger representation of women in tourism than other sectors include a lower emphasis on formal education and training, greater emphasis on personal and hospitality skills, higher availability of part time and work-from-home options, and opportunities for entrepreneurship that do not require substantial start-up investment (Twining-Ward & Zhou, 2017).

In some countries, tourism has almost twice as many women employers as other sectors, and the International Labour Organization (ILO) has found that women make up between 60 and 70 per cent of hotel labour force (ILO, 2010). Despite this level of representation, they are generally paid 10–15 per cent less than their male equivalents (UNWTO & UN Women, 2010). Women tend to dominate lower-paid jobs, such as clerical and cleaning roles, and are under-represented in higher-paid roles, such as tour guides, chefs, and particularly in management and decision-making positions (Twining-Ward & Zhou, 2017). Furthermore, women are more likely to be victims of sexual exploitation in tourism. In response, initiatives such as The Code (short for 'The Code of Conduct for the Protection of Children from Sexual Exploitation in Travel and Tourism') have arisen as multi-stakeholder responses to raise awareness and to provide tools and support to prevent abuse.[1]

In South Africa and Lesotho, work by the ILO has emphasized the empowerment of women in the tourism sector. Here, technical support has been provided to guest house and bed and breakfast establishments owned by women, to enhance the commercial viability of their enterprises (Rylance and Spenceley, 2013a; 2013b).

Goal 6. Clean water and sanitation: Ensure availability and sustainable management of water and sanitation for all

Tourism investment requirements for providing utilities can play a critical role in achieving water access and security, as well as hygiene and sanitation for all. The efficient use of water in tourism, pollution control and technological efficiency can be key to safeguarding water (UNWTO & UNDP, 2017).

As an example, tourism investment improving water access and security is provided by a community-owned facility managed by a private operator—Covane Lodge in Mozambique. One of the main challenges that the owners (the Canhane community) have is the lack of access to water. As part of a donor-funded infrastructure re-investment programme from the government of Mozambique's Mozbio project, a pipeline and pump system was installed from the Massingir dam to the community. To ensure sustainability, a community maintenance fund was established, so that people pay for use of the water (World Bank, 2014). Such initiatives can

1 See http://www.thecode.org/.

substantially improve the quality of life for the poor in dry areas, particularly for women, livestock owners and farmers.

Goal 7. Affordable and clean energy: Ensure access to affordable, reliable, sustainable and modern energy for all

Tourism can accelerate the shift towards increased use of renewable energy, and by promoting investments in clean energy sources the sector can help to reduce greenhouse gases, mitigate climate change and contribute to access of energy for all (UNWTO & UNDP, 2017).

While designing tourism infrastructure to integrate low energy use and installing renewable energy technologies is easiest at the outset, some facilities decide to retrofit their operations to make them more energy efficient. For example, Mombo camp in the Okavango Delta of Botswana switched from diesel generators to 100 per cent solar energy in May 2012, following a capital investment of approximately US$860,000. It was predicted that this change would lead to a 93 per cent reduction in carbon emissions from the camp, meaning that the lodge would emit only 22 tonnes of carbon dioxide equivalents during the year, compared to an estimated 287 tonnes during their 2012 financial year (Wilderness Holdings, 2014). In Ethiopia, Bale Mountain Lodge is 100 per cent eco-friendly with power coming from its own 25Kw micro-hydro power plant—biodegradable waste is processed through its biogas system to provide cooking gas and firewood is sourced from sustainable plantations outside the National Park (Snyman & Spenceley, 2019).

Goal 8. Decent work and economic growth: Promote sustained, inclusive and sustainable economic growth, full and productive employment and decent work for all

Tourism in protected areas represents one of the economic opportunities that can help to achieve the joint objectives of sustainable livelihood development of local communities and biodiversity conservation. Globally, tourism in 2015 accounted for 10 per cent of global gross domestic product (GDP), 7 per cent of global trade and one in ten jobs (UNWTO, 2017c). The flow of money from this sector provides opportunities for tourists to act as conduits to redistribute wealth from the rich to the poor. For example, when travellers visit developing countries they spend money on transport, accommodation, excursions, shopping, and on food and drink. Much of this money can be captured by local poor people if they are able to supply the products and services that tourists need, or if they are employed in tourism businesses (Spenceley & Meyer, 2012).

Although it is generally understood that tourism can contribute to protected area revenues, the significance of its actual and potential contribution is not well documented (Rathnayake, 2016). The United States Parks Service was one of the first protected area systems internationally to regularly determine their impact on the national economy, using a Money Generation Model. It estimated that its protected area system contributed to creating 251,600 jobs, US$9.34 billion in labour income, and US$16.5 billion in value addition to the national economy in 2012 (Cui, Mahoney & Herbowicz, 2013).

In terms of the economic impact of tourism on local communities, reports indicate that the benefits accrued by employees and host communities from tourism vary widely between enterprises and destinations, and are dependent on the institutional structures, partnership arrangements and business viability of a particular venture (Dedeke, 2017; Spenceley, 2008).

Inclusive business in tourism is a relatively new term, closely linked to the concept of pro-poor tourism (PPT). Although the principles of using tourism to reduce poverty, through 'sustainable', 'responsible', 'pro-poor' approaches and by strengthening 'value-chain linkages' are well established (see Spenceley & Meyer, 2012) the application of inclusive business to tourism is relatively new. Inclusive business in tourism is defined as including people living in poverty as customers, employees and entrepreneurs at various points along the value chain for mutual benefit (Tewes-Gradl, Van Gaalen & Pirzer, 2014). Inclusive business in tourism advocates a business-oriented approach, which aims to support sustainable development and poverty reduction through business partnerships. Inclusive business aims to encourage businesses to retain their commercial principles in investment decisions, but also incorporates their potential value to society into their decision-making. It requires innovation, and during periods of financial recession companies that are able to identify new opportunities, markets, supply chains and customers improve their resilience and increase their competitive advantage (Ashley, 2009).

Governments can take policy decisions and establish incentives to formalize and improve the standards of emerging businesses and entrepreneurs, so that more of the revenue from tourism can be captured and retained locally.

Goal 9. Industry, innovation and infrastructure: Build resilient infrastructure, promote inclusive and sustainable industrialization and foster innovation

Tourism development relies on good public and private infrastructure, and the sector can influence public policy for improvements that attract tourists and foreign investment (UNWTO & UNDP, 2017), while also supporting local communities.

Environmental impacts of infrastructure include water pollution, visual and sound disturbance, and invasive alien species. Environmental footprints extend beyond the infrastructure itself. Construction impacts include lighting, construction noise, vehicle movements, earth-moving operations, slopewash and turbid runoff from earthworks, water and air pollution, wastes, introduction of weed seeds and pathogens, and the introduction of feral animals. Large-scale visitor infrastructure can lead to habitat fragmentation, vehicular collisions with wildlife, traffic noise and light pollution, while new roads and visitor trails can lead to the spread of invasive alien species. New infrastructure increases visitation, creating further impacts and pressures for further site hardening (Leung et al., 2018).

Impact management approaches are reviewed by Buckley (2004; 2009; 2011; 2012), and can differ greatly in scale. Technologies for sewage and wastewater treatment, for example, may range from small-scale composting toilets for low-visitation infrastructure in warm moist climates, to multi-stage industrial sewage treatment systems with artificial wetland and ponds, appropriate for infrastructure with high visitor volumes. Controlling diffuse impacts is especially challenging.

Goal 10. Reduced inequalities: Reduce inequality within and among countries

Tourism can be a powerful tool for reducing inequalities if it engages local populations and all key stakeholders in its development, and can also contribute to urban renewal and rural development by giving people the opportunity to prosper in their place of origin (UNWTO & UNDP, 2017).

In low- and middle-income countries, tourism can generate substantial foreign exchange. Estimates suggest that tourism can contribute up to 40 per cent of GDP in less developed countries compared to 10 per cent of GDP in more economically advanced countries (Sofield et al., 2004). Many developing countries are rich in cultural and natural attractions, and their economies rely on tourism. For example, in 2005, tourism in Samoa accounted for 80 per cent of total goods and services exports (Meyer, 2010). These foreign exchange earnings from tourism are particularly important in developing countries where government strategies aim to reduce the proportion of their citizens living in poverty. It is estimated that approximately 20 per cent of the world's population (1.4 billion people) are currently living in 'extreme poverty' as they survive on less than US$1.25 per day (Chen & Rothschild, 2010; Wroughton, 2008).

This said, tourism also has the potential to negatively impact on communities from social, environmental and economic perspectives (Ashley, Boyd & Goodwin, 2000; Diaz, 2001; Koea, 1977). As a result, tourism has often been promoted as an opportunity to achieve both livelihood diversification and poverty reduction, while also acknowledging that it can generate negative impacts for the poor, such as displacement, price inflation of local products, and increased competition with which local community businesses may not be able to compete (Roe & Urquhart, 2002). For example, a survey of 17 marine protected areas (MPAs) in Thailand identified that local communities were believed to receive negligible benefits from tourism livelihoods (Bennett & Dearden, 2014). It is therefore important to consider the negative influences of tourism (or lack of benefits) alongside the positive benefits.

There is broad consensus that tourism is not a definitive answer to curing poverty but rather a tool within a larger portfolio of measures (Saarinen & Rogerson, 2014). This is because of the difficulty of local communities in accessing tourism value chains (Adiyia et al., 2015), as well as the high level of revenue leakage out of a particular destination (Britton, 1991). It is debatable whether the tourism industry considers any ethical commitment as a contribution towards global development objectives (Scheyvens, 2009).

Goal 11. Sustainable cities and communities: Make cities and human settlements inclusive, safe, resilient and sustainable

Over recent decades population sizes in urban areas have boomed and are expected to continue to grow, increasing by 61 per cent by 2030. It is predicted that the volume of people living in cities will rise to 5 billion by 2030 (UNWTO, 2012). The growth of the tourism sector can provide employment opportunities for these bulging populations. Tourism can help to advance urban infrastructure and accessibility, promote regeneration and preserve cultural and natural heritage, assets on which tourism depends. Furthermore, investment in green infrastructure (more efficient transport, reduced air pollution) should result in smarter and greener cities, not only for residents but also for tourists (UNWTO & UNDP, 2017).

Historical or cultural sites located in urban centres can be important economic drivers of a local economy, creating direct employment as well as strengthening linked value chains. Increased visitation can encourage preservation of national monuments, such as UNESCO World Heritage Sites. However, excessive levels of tourism can also have negative impacts, potentially damaging sites unless they are well managed. Furthermore, the capacity of city resources, such as water and electricity, can be overextended due to the demand from excessive tourism.

Goal 12. Responsible consumption and production: Ensure sustainable consumption and production patterns

The tourism sector can adopt sustainable consumption and production (SCP) modes, and so accelerate the shift towards sustainability (UNWTO & UNDP, 2017). The Sustainable Tourism Programme of the UNWTO-led Ten-Year Framework of Programmes (10YFP) on Sustainable Consumption and Production Patterns aims to catalyse change in tourism and promote transformation for sustainability (UNEP, n.d.). The programme aims to apply a life-cycle approach to development, through planning, investment, operations, production and consumption of sustainable goods and services, and monitoring and evaluation, among other areas (UNEP, n.d.).

For example, a programme on the Great Barrier Reef in Australia encourages tourism companies (e.g. accommodation providers, tour operators) working in protected areas to operate sustainably. The Marine Park Authority preferentially promotes operators that are independently certified by recognized environmental certification schemes, such as EarthCheck and Ecotourism Australia (GBRMPA, 2018). Where protected area authorities promote sustainable tourism in this way, they can promote more responsible consumption and production within MPAs.

Furthermore, protected area managers in Australia reward and encourage tour operators to become certified by offering longer licences, exclusive access to sensitive sites and promotional opportunities. These no-cost approaches demonstrate to operators that being sustainable, and independently certified as so, makes business sense (R. Hillman, chief executive, Ecotourism Australia, pers. comm., 11 April 2016). However, the level of uptake of certification in the tourism sector is low. A study undertaken in Africa established that there were only 715 hotels with sustainable tourism certifications offered by 18 certification bodies. Furthermore, these were present in only 18 countries on the continent. This means that only around 3 per cent of hotels in Africa have been independently established as operating sustainably (Spenceley, 2016). So, clearly, more needs to be done to mainstream monitoring and reporting of responsible consumption and production in the tourism sector.

Goal 13. Climate action: Take urgent action to combat climate change and its impacts

Tourism contributes to and is affected by climate change. By reducing its carbon footprint in the transport and accommodation sector, tourism can benefit from low carbon growth and help tackle one of the most pressing challenges of our time (UNWTO & UNDP, 2017).

The UNWTO states that governments and the private sector need to play an active role in responding to the challenges of climate change. Governments can act by establishing and implementing appropriate regulations and economic incentives. The private sector can commit to substantial reductions in the amount of carbon emissions they produce (UNWTO, 2009). The World Travel and Tourism Council facilitated a commitment by 40 of the world's largest travel and tourism companies to cut by half their greenhouse gas (GHG) emission levels of 2005 by 2035 (WTTC, 2009). The World Economic Forum suggests a series of options that the tourism industry could apply to mitigate GHG emissions that relate to land and air transport, water transport and accommodation. In essence, reductions in carbon emissions can be driven by the use of existing mature technologies in lighting, heating and cooling in order to improve hotel energy efficiency (Chisea & Gautam, 2009). Specific actions that tourism enterprises can undertake include hotel refurbishment to support the highest degree of energy efficient heating, cooling, lighting and building technology through incentives for energy efficient investments or mandatory energy efficiency certificates (Chiesa & Gautam, 2009).

Energy saving options for hotels include establishing environmental management systems, reducing energy use, using only renewable energy, reusing materials, recycling waste, using locally produced food and constructing low-carbon buildings from recycled materials with high levels of insulation (Simpson et al., 2008). The examples provided for SDG Goal 7, from Ethiopia and South Africa, illustrate how tourism businesses can reduce their greenhouse gas footprint. In Mauritius, the Association des Hôteliers et Restaurateurs–Ile Maurice collaborated with the

Commonwealth Secretariat to produce 'guidelines for energy efficiency' for the tourism sector. These included appliance selection tools (e.g. for air-conditioning, ovens, laundry equipment, lighting, solar photovoltaic, and water heating systems). Sample energy plans were provided for appliances to provide practical guidance to tourism operators (Spenceley, Bashain & Saini, 2011).

Goal 14. Life below water: Conserve and sustainably use the oceans, seas and marine resources for sustainable development

Since the 1960s there has been a ten-fold increase in the number of PAs (Emerton, Bishop & Thomas, 2006). In 2016, there were 202,467 terrestrial and inland water PAs recorded in the World Database on Protected Areas (WDPA), covering 14.7 per cent (19.8 million km^2) of the world's extent of ecosystems (excluding Antarctica). Consequently, to attain 17 per cent of terrestrial coverage outlined in Aichi Target 11, an additional 3.1 million km^2 would need to be protected. There are 14,688 Marine Protected Areas (MPA) recorded in the WDPA, covering 4.12 per cent (14.9 million km^2) of the global ocean and 10.2 per cent of coastal and marine areas under national jurisdiction (UNEP–WCMC & IUCN, 2016).

Coastal and maritime tourism rely on healthy marine ecosystems, and tourism development must help conserve and preserve fragile marine ecosystems and serve as a vehicle to promote a blue economy, contributing to the sustainable use of marine resources (UNWTO & UNDP, 2017). For example, research in Guam found that asking diving tourists to watch their buoyancy and avoid touching coral reefs led to a 75 per cent reduction in accidental contacts with the reefs, so reducing damage to these highly sensitive systems (Williams & Raymundo, 2017). In coastal areas, tourism facilities must ensure safe solid and liquid waste disposal, and improve the collection, safe disposal, and recycling of waste. For example, the Constance Ephelia resort in the Port Launay Marine Park of the Seychelles undertakes extensive recycling and waste recovery programmes, and is certified by two independent certification programmes (Green Globe, and the Seychelles Sustainable Tourism Label) (Spenceley, 2018c).

Goal 15. Life on land: Protect, restore and promote sustainable use of terrestrial ecosystems, sustainably manage forests

Rich biodiversity and natural heritage are often the main reasons why tourists visit a destination. Tourism can play a major role if sustainably managed in fragile zones, not only in conserving and preserving biodiversity, but also in generating revenue as an alternative livelihood for local communities (UNWTO & UNDP, 2017). The IUCN's World Commission on Protected Areas (WCPA) has a Tourism and Protected Areas Specialist Group (TAPAS Group), which has collated extensive work in this field. For example, IUCN Best Practice Guidelines on sustainable tourism (Eagles, McCool and Haynes, 2002), and tourism and visitor management in protected areas (Leung et al., 2018) collated extensive international experiences on tourism in protected areas.

In Africa, Snyman and Spenceley (2019) have compiled a series of case studies on the conservation benefits of tourism facilities in protected areas. For example, Nkwichi Lodge in Mozambique, created a 120,000 hectare new community conservation area that coordinated efforts of 16 communities to help regulate land use and, in particular, to stop hunting. In Rwanda, Bisate Lodge has established an extensive reforestation programme. Using seed and other material gathered only from outside the park, an indigenous tree nursery was established and more than 15,000 trees were planted between 2015 and 2017. Trees have also been donated to reforest other areas around Volcanoes National Park (Snyman & Spenceley, 2019).

Goal 16. Peace, justice and strong institutions: Promote peaceful and inclusive societies for sustainable development, provide access to justice for all and build effective, accountable and inclusive institutions at all levels

As tourism revolves around billions of encounters between people of diverse cultural backgrounds, the sector can foster multicultural and inter-faith tolerance and understanding, laying the foundation for more peaceful societies. Tourism, which benefits and engages local communities, can also consolidate peace in post-conflict societies (UNWTO & UNDP, 2017).

Transfrontier Conservation Areas (TFCAs) are described as relatively large areas encompassing one or more protected areas which straddle frontiers between one or more countries (World Bank, 1996), and are sometimes called 'peace parks'. By contrast to national parks, TFCAs have the potential to conserve a greater diversity of species within larger geographical areas and to promote cooperative wildlife management between countries (BSP, 1999). TFCAs may also improve opportunities for tourism by allowing visitors to disperse over greater areas and obtain better quality experiences (Singh, 1999). In southern Africa, there have been extensive initiatives to establish transboundary tourism that support inclusive approaches to tourism across international borders. These include the development of guidelines for tourism concession in TFCAs (Spenceley, 2014) and also for cross-border tourism products (Spenceley, 2018a). These tools seek to establish transparent and well-governed processes for tourism investment in protected areas that benefit conservation and host communities.

Goal 17. Partnerships for the goals: Strengthen the means of implementation and revitalize the global partnership for sustainable development

Due to its cross-sectoral nature, tourism has the ability to strengthen private/public partnerships and engage multiple stakeholders—international, national, regional and local—to work together to achieve the SDGs and other common goals (UNWTO & UNDP, 2017).

In protected areas that are attractive to tourists, and which present a commercially viable opportunity, authorities have been able to establish tourism concessions, or public/private partnerships. These can be used by protected area authorities to spread the commercial risk associated with high value capital investment for infrastructure (e.g. luxury lodges). In some instances, these partnerships can contribute meaningful revenue to protected area budgets, which in turn contribute towards funding for conservation management (Spenceley, Snyman & Eagles, 2017). Funding for tourism partnerships, including joint-ventures between private operators and rural communities, can be sourced from development banks, aid agencies, impact investment vehicles, non-governmental organizations, government grants, and private investment (Snyman & Spenceley, 2019).

In Australia, the Great Barrier Reef Authority has a partnership with tour operators to collect a marine conservation fee (an Environmental Management Charge). This fee has been used to help finance conservation of the MPA. Although when introduced, the fee faced some controversy, following a decade of use and adaptation, the fee is well established and widely accepted (Skeat & Skeat, 2007).

Research challenges

Relative to the SDGs, there is a growing concern regarding the local-level impact on poverty alleviation and the attainment of the SDGs (Rogerson, 2006; Saarinen & Rogerson, 2014), as well as the effectiveness of using tourism to deliver economic development and conservation objectives (Bookbinder et al., 1998; Goodwin & Santilli, 2009; Shibia, 2010; Walpole, Goodwin & Ward, 2001).

Eagles (2014) suggests that based on the body of research work on protected areas and tourism that exists, there are ten key areas for protected area tourism research in the future. These are (1) visitor use monitoring; (2) park tourism economic impact monitoring; (3) park finance; (4) professional competencies for tourism management; (5) building public support; (6) visitor satisfaction; (7) licences, permits, leases and concessions for tourism; (8) pricing policies; (9) management capacity; and (10) park tourism governance.

Leung et al. (2018) highlight critical issues that need to be considered in tourism management of conservation areas in the future, which include: (1) population growth and increasing consumption, and increasing demands on tourism destinations; (2) urbanization, and the needs of residents for recreational experiences; (3) demographic changes with the rise of the middle class in highly populated countries (e.g. China and India's booming travel sectors); and (4) climate change, with extreme weather events increasing in frequency and intensity.

In its review of tourism and SDGs, UNWTO outlines a series of recommendations. These include strengthening cooperation and multi-stakeholder partnerships; designing and implementing incentives and smart subsidies; raising awareness of business opportunities created by the SDGs, and aligning development cooperation programmes with the needs and priorities of developing countries; and sharing experiences, good practices and lessons learned (UNWTO, 2017a).

The Global Sustainable Tourism Council is an independent non-profit organization that establishes and manages global standards, namely the GSTC Criteria. These are put forward as guiding principles and minimum requirements that the tourism sector should aspire to achieve in order to protect and sustain the world's natural and cultural resources, while contributing to conservation and poverty reduction (GSTC, 2017). The GSTC has mapped its Destination Criteria against the SDGs, and further demonstrates that sustainable tourism can comprehensively address them.[2] This means that certification programmes that have standards recognized by the GSTC (indicating their alignment with the GSTC Criteria) are evaluating tourism destinations against all of the SDGs. So, effectively, it means that the tourism destinations certified using the GSTC-recognized standard are also implementing the 17 SDGs through their visitor management approaches. However, what is missing is the ability for us to tap into the data collected by certification bodies globally, and collate these achievements on the SDGs.

In June 2018, the government of Mauritius hosted an international conference on Sustainable Tourism and Digitization, where presentations were made on sustainable tourism. One of the questions raised was how researchers can use information technology to help nations monitor and report on the SDGs. A wealth of qualitative and quantitative data is embedded within case studies, technical reports and studies that address one or two, but not all, of the SDGs. How can all the data be collated and assimilated to demonstrate tourism's achievement towards the SDGs? (Spenceley, 2018b). More research in this arena is certainly required to capture achievements and understand the practical challenges of achieving sustainable development.

Summary

Tourism's capacity to contribute to sustainable development if managed effectively was recognized in 2017 by the UN Year of Sustainable Tourism for Development. The analysis above has highlighted how tourism can be applied across the 17 Global Goals for the year 2030, and can be seen as a major tool in accomplishing these targets. This chapter has highlighted the linkages between the tourism sector and SDGs that have a clear economic (SDG 1, 2, 8, 9, 10, 11, 12) or environmental focus (SDG 6, 7, 13, 14, 15), but the impact of the sector on societal goals (SDG 3, 4, 5, 16) are less immediately obvious. This does not mean that they are absent, but simply the direct attribution of the tourism sector is less clear. Examples of how tourism businesses have implicitly contributed towards these Global Goals are provided for each SDG.

2 See https://www.gstcouncil.org/sdgs-gstc-destination-criteria/.

Tourism has the potential to positively contribute towards every SDG, but it also has the potential to increase inequality, exacerbate economic leakages, and over-consume water resources. How the sector is managed often determines the outcome—the legal framework to set the foundation and provide the direction, the incentives to behave well, the penalties for those who choose not to, as well as, most importantly, the overriding intentions of the tourism operators and market demand from their customers. Managing a sector to achieve development goals simultaneously, and identifying and addressing conflicting goals quickly, such as visitor growth versus water availability, requires an appreciation of complexity and innovation. This chapter has highlighted a number of examples of good practice by individual operators, groups or destinations. But this is by no means systemic. The evidence for a business case for sustainable tourism continues to grow but more rapid mainstreaming is required in practice.

References

Adiyia, B., Stoffelena, A., Jennesa, B., Vanneste, D. & Ahebwa, W. (2015). 'Analysing governance in tourism value chains to reshape the tourist bubble in developing countries: the case of cultural tourism in Uganda'. *Journal of Ecotourism*, **14**(2–3): 111–29.

Ashley, C. (2009). *Harnessing Core Business for Development Impact: Evolving Ideas and Issues for Action*. London: Overseas Development Institute.

Ashley, C., Boyd, C. & Goodwin, H. (2000). *Pro-Poor Tourism: Putting Poverty at the Heart of the Tourism Agenda*. Natural Resource Perspectives No. 51. London: Overseas Development Institute.

Ashley, C., Mdoe, N. & Reynolds, L. (2002). 'Rethinking wildlife for livelihoods and diversification in rural Tanzania: a case study from northern Selous'. Working Paper No. 15. London: Overseas Development Institute.

Ashley, C. & Mitchell, J. (2007). *Assessing How Tourism Revenues Reach the Poor* (ODI Briefing Paper No. 21). London: Overseas Development Institute.

Ashley, C., Roe, D. & Goodwin, H. (2001). *Pro-poor tourism strategies. Making tourism work for the poor: a review of experience*. Pro-poor tourism report No. 1, April 2001, ODI/IIED/CRT, The Russell Press.

Bennett, N. & Dearden, P. (2014). 'Why local people do not support conservation: community perceptions of marine protected area livelihood impacts, governance and management in Thailand'. *Marine Policy*, **44**: 107–16.

Bennett, O., Ashley, C. & Roe, D. (1999). *Sustainable Tourism and Poverty Elimination Study: A Report to the Department for International Development*. London: Deloitte & Touche.

Bookbinder, M., Dinerstein, E., Rijal, A., Cauley, H. & Rajouria, A. (1998). 'Ecotourism's support of biodiversity conservation'. *Conservation Biology*, **12**: 1399–1404.

Braack, J. and Mearns, K. (2017). 'Oceans without borders', presentation at the Conference on Sustainable Tourism in Small Island Developing States, 23–4 November 2017, University of Seychelles, Mahe, The Seychelles.

Britton, S. (1991). 'Tourism, capital, and place: towards a critical geography of tourism'. *Environment and Planning D: Society & Space*, **9**(4): 451–78.

BSP (Biodiveristy Support Program) (1999). 'Study on the development of transboundary natural resource management areas in southern Africa: highlights and findings'. Biodiversity Support Program, Washington DC: World Wildlife Fund. Cited in A. Spenceley (2007), 'Tourism in the Great Limpopo Transfrontier Park', *Development Southern Africa*, **23**(5): 649–69.

Buckley, R. C. (ed.) (2004). *Environmental Impacts of Ecotourism.* Wallingford: CABI.

Buckley, R. C. (2009). *Ecotourism: Principles and Practices.* Wallingford: CABI.

Buckley, R. C. (2011). 'Tourism and environment'. *Annual Review of Environment and Resources,* **36**: 397–416.

Buckley, R. C. (2012). 'Sustainable tourism: research and reality'. *Annals of Tourism Research,* **39**(2): 528–46. Available at https://doi.org/10.1126/science.344.6182.358-b.

Chen, C. & Rothschild, R. (2010). 'An application of hedonic pricing analysis to the case of hotel rooms in Taipei'. *Tourism Economics,* **16**(3): 685–94.

Chiesa, T. & Gautam, A. (2009). 'Towards a low carbon travel and tourism sector' (with Booz & Company). Geneva: World Economic Forum.

Cui, Y., Mahoney, E. & Herbowicz, T. (2013). 'Economic benefits to local communities from national park visitation'. Natural Resource Report NPS/NRSS/EQD/NRTR—2013/631. Colorado: US National Park Service.

Dedeke, A. (2017). 'Creating sustainable tourism ventures in protected areas: an actor-network theory analysis'. *Tourism Management,* **61**: 161–72.

Department of National Parks (2014). 'Strategic plan 2015–2019'. Queensland Government.

Diaz, D. (2001). 'The viability and sustainability of international tourism in developing countries'. Report to the Symposium on Tourism Services. Geneva: World Trade Organisation.

Eagles, P. F. J. (2014). 'Research priorities in park tourism'. *Journal of Sustainable Tourism,* **22**(4): 528–49.

Eagles, P. F. J., McCool, S. F. & Haynes, C. (2002). 'Sustainable tourism in protected areas: guidelines for planning and management'. Best Practice Protected Area Guidelines Series No. 8. Gland, Switzerland: IUCN.

Emerton, L., Bishop, J. & Thomas, L. (2006). *Sustainable Financing of Protected Areas: A Global Review of Challenges and Options.* Gland, Switzerland and Cambridge, UK: IUCN.

Goodwin, H. & Santilli, R. (2009). *Community-Based Tourism: A Success?* ICRT Occasional Paper, 11. Greenwich: ICRT & GTZ.

Great Barrier Reef Marine Park Authority (GBRMPA) (2018). 'Choosing a high standard tourism operation'. Available at http://www.gbrmpa.gov.au/visit-the-reef/choose-a-high-standard-operator.

GSTC (Global Sustainable Tourism Council) (2017). 'What is the GSTC?' Available at https://www.gstcouncil.org/about/about-us/ (accessed 2 August 2018).

HPHP (Healthy Parks Healthy People) (2017). 'Healthy parks healthy people central'. Available at http://www.hphpcentral.com (accessed 15 February 2017).

ILO (International Labour Organization) (2010). 'Developments and challenges in the hospitality and tourism sector – issues paper for discussion at the Global Dialogue Forum, for Hotels, Catering, and Tourism'. Geneva: International Labour Organization.

Koea, A. (1977). 'Polynesian migration to New Zealand'. In B. Finney & A. Watson (eds), *A New Kind of Sugar: Tourism in the Pacific,* pp. 69–9. Santa Cruz: Centre for South Pacific Studies, University of California.

Leung, Y.-F., Spenceley, A., Hvenegaard, G. & Buckley, R. (2018). 'Tourism and visitor management in protected areas: guidelines for sustainability'. Best Practice Protected Area Guideline Series No. 27. Geneva: IUCN.

Meyer, D. (2010). 'Pro-poor tourism: can tourism contribute to poverty reduction in less economically developed countries?' In S. Cole & N. Morgan (eds), *Tourism and Inequality: Problems and Prospects,* pp. 164–82. London: CABI.

Mitchell, J. & Ashley, C. (2009). *Value Chain Analysis and Poverty Reduction at Scale* (ODI Working Paper No. 49). London: Overseas Development Institute.

Naughton-Treves, L., Holland, M. & Brandon, K. (2005). 'The role of protected areas in conserving biodiversity and sustaining local livelihoods'. *Annual Review of Environment and Resources,* **30**: 219–52.

Rathnayake, R. (2016). 'Economic values for recreational planning at Horton Plains National Park, Sri Lanka'. *Tourism Geographies,* **18**(2): 213–32.

Roe, D. & Urquhart, P. (2002). *Pro-Poor Tourism: Harnessing the World's Largest Industry for the World's Poor.* London: IIED.

Rogerson, C. (2006). 'Pro-poor local economic development in South Africa: the role of pro-poor tourism'. *Local Environment,* **11**(1): 37–60.

Rylance, A. & Spenceley, A. (2013a). 'Having the perseverance and confidence in yourself to make the right business decisions, Case Study: Mantovani Guest House', Report to the ILO.

Rylance, A. & Spenceley A. (2013b). 'Female-owned tourism businesses in Lesotho, Case Study: Mokhotlong Hotel and Motlejoa Guest House', Report to the ILO.

Saarinen, J. & Rogerson, C. (2014). 'Tourism and the Millennium Development Goals: perspectives beyond 2015'. *Tourism Geographies,* **16**(1): 23–30.

Scheyvens, R. (2009). 'Pro-poor tourism: is there value beyond the rhetoric?' *Tourism Recreation Research,* **34**(2): 191–6.

Scheyvens, R. (2011). *Tourism and Poverty.* New York: Routledge.

Schiere, H. & Kater, L. (2001). *Mixed Crop- Livestock Farming: A Review of Traditional Technologies on Literature and Field Experience.* Rome: FAO.

Sen, A. (1999). *Development as Freedom.* Oxford: Oxford University Press.

Shibia, M. (2010). 'Determinants of attitudes and perceptions on resource use and management of Marsabit National Reserve, Kenya'. *Journal of Human Ecology,* **30**(1): 55–62.

Simpson, M. C., Gössling, S., Scott, D., Hall, C. M. & Gladin, E. (2008). *Climate Change Adaptation and Mitigation in the Tourism Sector: Frameworks, Tools and Practices.* Paris: UNEP, University of Oxford, UNWTO, WMO.

Singh, J. (1999). 'Study on the development of transboundary natural resource management areas in southern Africa – global review: lessons learned'. Biodiversity Support Program, Reference No. 59. Washington, DC: World Wildlife Fund. Cited in A. Spenceley (2007), 'Tourism in the Great Limpopo Transfrontier Park', *Development Southern Africa,* **23**(5): 649–69.

Skeat, A. & Skeat, H. (2007). 'Tourism on the Great Barrier Reef: a partnership approach'. In R. Bushell & P. F. J. Eagles (eds), *Tourism and Protected Areas: Benefits Beyond Boundaries: The Vth IUCN World Parks Congress,* pp. 315–28. Wallingford: CABI.

Snyman, S. & Spenceley, A. (2019). *Private Sector Tourism in Conservation Areas in Africa.* Wallingford: CABI.

Sofield, T., De Lacy, T., Lipman, G. & Daughety, S. (2004). *Sustainable Tourism – Eliminating Poverty (ST–EP): An Overview.* Tasmania: Australian Cooperative Research Centre for Sustainable Tourism.

Sparkes, C. & Woods, C. (2009). *Linking People to Landscape: The Benefit of Sustainable Travel in Countryside Recreation and Tourism.* N.p.: East of England Development Agency. Cited in Y.-F. Leung, A. Spenceley, G. Hvenegaard & R. Buckley (eds) (2018), *Tourism and Visitor Management in Protected Areas: Guidelines for Sustainability.* Best Practice Protected Area Guidelines Series No. 27. Gland: IUCN.

Spenceley, A. (2008). 'Impacts of wildlife tourism on rural livelihoods in Southern Africa'. In Spenceley (ed.), *Responsible Tourism: Critical Issues for Conservation and Development,* pp. 159–86. Sterling, VA: Earthscan.

Spenceley, A. (2014). 'Tourism concession guidelines for Transfrontier conservation areas in SADC'. Report to GIZ/SADC.

Spenceley, A. (2016). 'Green certification in the tourism sector in Africa: monitoring water and waste'. Report to the African Development Bank.

Spenceley, A. (2017). 'Tourism and protected areas: comparing the 2003 and 2014 IUCN World Parks Congress'. *Tourism and Hospitality Research,* **17**(1): pp. 8–23.

Spenceley, A. (2018a). 'SADC guidelines for cross-border tourism products'. Report to GIZ, 26 March.

Spenceley, A. (2018b). 'Tourism in Africa and the Sustainable Development Goals', Presentation at the International Conference on Digitalisation and Sustainable Tourism, May. Le Meridien, Ponte au Piments, Mauritius.

Spenceley, A. (2018c). 'Sustainable tourism certification in the African hotel sector'. *Tourism Review*. Available at https://www.emeraldinsight.com/doi/full/10.1108/TR-09-2017-0145.

Spenceley, A., Bashain, A. & Saini, A. (2011). 'Design of environmental good practice guidelines for the Mauritius hotel industry', Handbook and Implementation Plan, Project reference: X/MRT/027.

Spenceley, A. & Meyer, D. (2012). 'Tourism and poverty reduction: theory and practice in less economically developed countries'. *Journal of Sustainable Tourism*, **20**(3): 297–317.

Spenceley, A. & Meyer, D. (2016). *Tourism and Poverty Reduction: Principles and Impacts in Developing Countries*. London and New York: Routledge.

Spenceley, A., Snyman, S. & Eagles, P. (2017). *Guidelines for Tourism Partnerships and Concessions for Protected Areas: Generating Sustainable Revenues for Conservation and Development*. Report to the CBD. Available at https://www.cbd.int/tourism/doc/tourism-partnerships-protected-areas-web.pdf.

Tewes-Gradl, C., Van Gaalen, M. & Pirzer, C. (2014). *Destination Mutual Benefit: A Guide to Inclusive Business in Tourism*. Frankfurt: GIZ.

Twining-Ward, L. & Zhou, V. (2017). 'Tourism for development: women and tourism: designing for inclusion'. Washington, DC: World Bank.

UNEP (n.d.). '10 YFP Sustainable Tourism Programme'. Available at http://web.unep.org/10yfp/programmes/sustainable-tourism-programme.

UNEP–WCMC & IUCN (United Nations Environment Programme–World Conservation Monitoring Centre & International Union for Conservation of Nature) (2016). *Protected Planet Report 2016*. Cambridge, UK and Gland, Switzerland: UNEP-WCMC and IUCN.

UNWTO (United Nations World Tourism Organization) (2009). 'From Davos to Copenhagen and beyond: advancing tourism's response to climate change'. UNWTO Background Paper. Madrid: UNWTO.

UNWTO (United Nations World Tourism Organization) (2012). 'Global report on city tourism—cities 2012 project'. AM Report No. 6. Madrid: UNWTO.

UNWTO (United Nations World Tourism Organization) (2017a). 'Tourism and the Sustainable Development Goals – Journey to 2030, Highlights'. Available at http://publications.unwto.org/publication/tourism-and-sustainable-development-goals-journey-2030 (accessed 14 April 2018).

UNWTO (United Nations World Tourism Organization) (2017b). 'UNWTO Tourism Highlights, 2017 Edition'. Available at https://www.e-unwto.org/doi/book/10.18111/9789284419029 (accessed 14 April 2018).

UNWTO (United Nations World Tourism Organization) (2017c). *UNWTO Annual Report 2016*. Madrid: UNWTO.

UNWTO & UN Women (United Nations World Tourism Organization & UN Women) (2010). 'Global report on women in tourism'. Madrid: UNWTO.

UNWTO & UNDP (United Nations World Tourism Organization and United Nations Development Programme) (2017). *Tourism and the Sustainable Development Goals – Journey to 2030*. Madrid: UNWTO.

Walpole, M., Goodwin, H. & Ward, K. (2001). 'Pricing policy for tourism in protected areas: lessons from Komodo National Park, Indonesia'. *Conservation Biology*, **15**: 218–27.

Wilderness Holdings (2014). *Integrated Annual Report for the Year Ended 28 February 2014*. Wilderness Holdings, South Africa and Botswana.

Williams, A. & Raymundo, L. (2017). 'Ask and you shall receive: reducing diver impacts on Guam's coral reefs with a coral-safe diving reminder', presentation at the Sustainable Tourism in SIDS conference, 24–5 November, The Seychelles.

World Bank (1996). 'Mozambique, transfrontier conservation areas pilot and institutional strengthening project'. Maputo: Global Environment Facility Project Document. Cited in A. Spenceley (2007), 'Tourism in the Great Limpopo Transfrontier Park', *Development Southern Africa*, **23**(5): 649–69.

World Bank (2014). 'Mozambique, Transfrontier Conservation Areas and Tourism Development Project (TFCATDP) project implementation support mission', 4–14 February, aide memoire.

World Bank (2016a). 'Taking on inequality, poverty and shared prosperity'. Available at http://www.worldbank.org/en/publication/poverty-and-shared-prosperity (accessed 14 April 2018).

World Bank (2016b). *Gender Strategy, 2016–2023: Gender Equality, Poverty Reduction and Inclusive Growth*. Washington, DC: World Bank Group. Available at http://documents.worldbank.org/curated/en/820851467992505410/pdf/102114-REVISED-PUBLIC-WBG-Gender-Strategy.pdf.

World Tourism Organization (UNWTO) (2015). *Tourism and the Sustainable Development Goals*. Madrid: UNWTO.

Wroughton, L. (2008). 'More people living below poverty line'. World Bank, *Reuters*, Tuesday 26 August. Available at http://www.reuters.com/article/2008/08/26/idUSN26384266 (accessed 28 December 2011).

WTTC (World Travel and Tourism Council) (2009). 'Leading the challenge on climate change'. Available at http://www.wttc.org/eng/Tourism_Initiatives/Environment_Initiative/.

9 Business models and sustainable tourism

Kelly S. Bricker

Introduction

Tourism businesses generally comprise three types: (1) primary trades, commonly associated with the provision of tourism (e.g., tour operators, accommodation, transport, attractions); (2) secondary trades, those supportive but not exclusive to tourism (e.g., financial institutions, retail, entertainment, personal services); and (3) the tertiary trades, which provide basic infrastructure in a supportive role to tourism (e.g., public goods and services, manufacturing and food supply) (Berno & Bricker, 2001; Likorish & Jenkins, 1997). As such, despite the significant economic contribution to gross domestic product, tourism businesses remain significantly different from businesses in "other" industries, as they are interwoven into a series of complex systems involving resources within and beyond their immediate capacity to control. For example, Berno & Bricker (2001), in their critique of sustainable tourism as sustainable development, have raised several issues that may in fact be contrary to sustainable tourism businesses. They identified key challenges to operationalizing sustainable tourism as part of sustainable development, including the *nature of the tourism product, organizational fragmentation of the industry,* and questions surrounding *what is to be sustained.*

The nature of tourism

Tourism does not comprise the sale of a "common product." Rather it is intangible, and is comprised of the tourism experience (macro level product), or what the tourist actually "sees, uses, and experiences as part of their tourist encounter"; the place product, or the destination, a point of consumption and certain components that are tied directly to the tourist experience (i.e., a national park); and, of course, the tourism products, which refers to individual businesses/sectors in tourism, such as accommodation, attractions, restaurants, and souvenirs (Berno & Bricker, 2001, p. 6; see also O'Fallon, 1994). They further suggest tourism as a very different business product, as they include *tangible* (e.g., lodging, air tickets, horseback rides) and *intangible* (e.g., scenery, attitudes of local residents, sounds of the ocean) elements.

Products also have the opportunity for multi-consumption (more than one individual at a time, groups of people at a time), yet individual consumers of tourism products may perceive they are experiencing something unique and different, based on their engagement and so on. Further, they explain that "because of the intangible and service nature of many of the tourism products, they cannot be inspected prior to purchase or consumption. Once consumed the product is gone, and cannot be returned if 'damaged' or the 'incorrect fit'" (Berno & Bricker, 2001, p. 7). The immobility of a tourism product and the perishable nature of the tourism product (the experience, day and time disappear in the moment) also add increased complexity. Tourism products are also complimentary (e.g., transportation links to a destination) and tourists typically bundle these services together to form their overall experience or vacation. It is less typical that various products can survive without the other, demonstrating exhaustive linkages within the tourism system (Berno & Bricker, 2001). Notably, demand of tourism products is subject to externalities, out of direct control of the industry, such as political instability, natural disasters and foreign exchange rates (Berno & Bricker, 2001; O'Fallon, 1994), which leads us to the difficulties surrounding the organization of fragmentation.

Organizational fragmentation

As Leiper (2008) has stated, the "theory that all tourism falls within the ambit of one large industry is a false theory" (p. 249). Tourism's complexity is further evidenced in its organizational structure. From the individual, to community, to region, nation and planet, tourism is organized through various scales, public and political boundaries, with decisions made at various levels, each having the potential to stray from sustainable development, competing for resources that may impact the ability to support tourism and its associated infrastructure and environments. Decisions are often made in isolation at each level, with very little coordination at local, national, regional and international levels. As Berno & Bricker (2001) suggested, "sustainable tourism is in many ways about the competition for and distribution of finite resources, so in this respect, requires a political solution. A balance must be struck between tourism and other existing and potential activities" (p. 12). Leiper notes that "a highly developed tourism industry can be thought of as a chain with multiple links" (Leiper, 2004, pp. 150–51; 191–221).

In addition, the distribution of power plays an important role, which influences the nature of development within each sector (e.g., transportation, accommodation, tour operations). Also, diverse stakeholders in tourism have a multitude of rights and responsibilities within the tourism system. As Swarbrooke (1999) identified, groups interested in sustainable tourism have vastly different missions and responsibilities, from public sector (local, national, to multilateral bodies) each with distinctive priorities and level of influence on businesses, to tourism industries (complexities discussed above), voluntary sector organizations (such as centers, professional associations, and a range of pressure groups), host destinations and communities, media, and of course traveler (Berno & Bricker, 2001; Swarbrooke,

1999). Hence collaboration to solve transcendent issues facing planet and society make sense—and a rethinking of the business model.

The difficulties of integrating sustainable tourism into these various scales and goals, as well as the demand and supply sides of tourism, defy simplistic resolution (Berno & Bricker, 2001; Milne, 1998; see also this volume, Bosak (Chapter 2) and Bosak & McCool (Chapter 15)). As such, tourism businesses are the nexus of supply and demand, and significantly depend upon tourism resources within a destination (e.g., natural and cultural heritage). We now understand in tourism and other businesses that "theories based on linear frameworks that seek to understand cause and effect actions fail to account for multiple variables that affect businesses, the variation of contexts in which they operate, and the effect of temporal influences on all those factors" (Sun, Wu & Yang, 2018, p. 59). Therefore, an interesting dynamic is created which calls upon businesses to integrate rather than isolate for a long view of sustainable development and economic gain. How tourism businesses contribute to the larger system, react, and behave, can make important contributions to understanding complexity in sustainable tourism development.

While sustainable tourism remains operationally challenging, there are encouraging signs and demonstrations that tourism industries are achieving sustainable development and managing the challenges of "wicked" problems facing the planet today (One Planet Network, 2018). For example, the United Nations World Tourism Organization initiated the One Planet Network to refocus attention on sustainable consumption and production (SCP) within tourism's value chain. The mission of the sustainable tourism program is to "catalyze a transformation for sustainability, through evidence-based decision making, efficiency, innovation, collaboration among stakeholders, monitoring and the adoption of a life-cycle approach for continuous improvement" (One Planet Network, 2018, p. 1). This initiative recognizes tourism as a critical sector in improving sustainable development, justified by tourism's "transversal nature" and its complex linkages across related industries and supply chains (One Planet Network, 2018, p. 1).

One Planet considers the United Nations Sustainable Development Goals, in particular SDG 12 which focuses on ensuring SCP and includes "targets related to the sustainable management and efficient use of natural resources, food loss, waste generation, sustainability reporting of business and the development and implementation of tools to monitor sustainable development impacts of sustainable tourism . . ." (One Planet Network, 2018, p. 1). This initiative and many others contribute to global awareness and priority setting. Yet a deeper understanding of the functionality and operationalization of sustainability at the business level can assist in implementation of global initiatives and sustainable development goals.

Budeneau et al. (2015) suggest it is time to look outside tourism literature for concepts and models found to be of value in other disciplines. For example, tourism researchers could look to the green economy concept for building a strategic planning process model in tourism planning (Law et al., 2016), clusters and networks to explore organizational level concepts (McLennan, Becken & Watt, 2016), or examination of behavior change programs in public health to explore behavior change in organizations or tourists (Higham et al., 2016). Broadening perspective and models beyond tourism industries (e.g., operations, marketing, impact) and into a systems approach to understand tourism's placement into a broader social and environmental context, is critical if we are to understand sustainability in the context of this large and complex enterprise (Bricker, 2018).

The journey towards . . .

Early concerns that environmentally friendly business models would erode competitiveness and add to costs, without immediate financial gain, are still prevalent in many corporations (Nidumolu, Prahalad & Rangaswami, 2009). Yet research has demonstrated that these concerns appear to be unfounded (Esty & Winston, 2006; Nidumolu, Prahalad & Rangaswami, 2009). Research conducted on large corporations has demonstrated that sustainability and technological innovation can actually enhance "bottom-line and top-line returns" (Nidumolu, Prahalad & Rangaswami, 2009, p. 57). For companies adopting environmentally friendly models, costs are lowered due to a significant reduction of inputs utilized, opportunities are created for better products (resulting in additional revenues) and new innovations are enabled (Nidumolu, Prahalad & Rangaswami, 2009).

Nidumolu, Prahalad & Rangaswami identified five distinct stages of change which companies engage in on their sustainability journey. These stages are briefly summarized here for purposes of considering the application to tourism, including the challenges, competencies needed and opportunities afforded:

Stage 1: Viewing compliance as opportunity
Legal environmental standards vary by country, state, region and possibly city. In addition, companies often feel pressured to conform to voluntary codes (e.g., tourism: the Global Sustainable Tourism Criteria; wood: Forestry Stewardship Council; fisheries: Marine Stewardship Council). Stage 1 is an opportunity to address "emerging norms" and experiment early on with alternatives within product development. Furthermore, advantages of taking the lead include turning regulators into allies, and allowing time to team up with other like-minded companies to create solution efficiencies.

Challenge: to ensure that compliance with norms becomes an opportunity for innovation.
Competencies: the ability to anticipate and shape regulations.
Innovation opportunity: using compliance to induce a company and its partners to experiment with sustainable technologies, materials and processes.

Stage 2: Making value chains sustainable
Companies become more familiar with regulations and voluntary codes and begin to focus outwardly on suppliers, particularly regarding raw materials and

renewable energy, with a focus on creating a better image, reducing costs and creating new business. In addition, corporations offer incentives within their supply chain to induce suppliers to become environmentally sustainable. Other strategies have included operational innovations which enhance energy efficiency and reduce dependence on fossil fuels.

Challenge: to increase efficiencies throughout the value chain.
Competencies: expertise in carbon management, and life-cycle assessment; an ability to redesign operations for efficiency and reduction of water, waste and fossil fuels; capacity to ensure suppliers and retailers make their operations sustainable/eco-friendly.
Innovation opportunity: developing sustainable sources of materials and services; increasing the use of "clean" energy sources; life-cycle considerations of retired products.

Stage 3: Designing sustainable products and services
Companies acknowledge consumer preference for eco-friendly products and acknowledge a competitive advantage when re-designing or developing new products. It is critical for companies to identify consumer concerns, partner with relevant organizations that can assist with developing competencies beyond their traditional expertise, and look beyond public relations benefits in order to identify and increase competencies to enable future market leadership.

Challenge: to develop sustainable offerings or redesign existing ones to become eco-friendly.
Competencies: skills to understand products and services unfriendly to the environment; ability to generate public support for sustainable offerings, and not be considered "green washing"; require management know-how to scale supplies of materials and so on to create a green product.
Innovation opportunity: apply techniques such as biomimicry in product development; develop compact and eco-friendly packaging and sourcing.

Stage 4: Developing new business models
Critical to this success is the exploration of alternatives to current business practices, and finding innovation in new delivery mechanisms, including working in tandem with other companies.

Challenge: to find novel ways of capturing and delivering value, which change the basis of competition.
Competencies: a capacity to understand consumers and to figure out how to meet demands; an ability to understand how partners can enhance the value of offerings.
Innovation opportunity: develop new delivery technologies that change value-chain relationships in significant ways; create monetization models that relate to services rather than products; devise business models that combine digital and physical infrastructure.

Stage 5: Creating next-practice platforms
The final stage of change experienced by companies and their CEOs involves the idea that "sustainability can lead to interesting next practice" (Nidumolu, Prahalad & Rangaswami, 2009, p. 11). The authors' research reveals two enterprise-wide initiatives that enable companies to become more sustainable. First, a top management team focuses on a problem and change happens. Second, recruitment and retention of those who believe in social responsibility and a commitment to protecting the environment is essential. The bottom line, they suggest, is that sustainability = innovation.

Challenge: to question through a sustainability lens the dominant logic behind business today.
Competencies: knowledge of how renewable and non-renewable resources affect business ecosystems and industries; expertise in synthesizing business models, technologies and regulations in different industries.
Innovation opportunity: building platforms that will enable customers and suppliers to manage energy in radically different ways; developing products that reduce the need for water; designing technologies that allow industries to use energy produced as a by-product.

(For a detailed explanation of these stages, see Nidumolu, Prahalad & Rangaswami, 2009, pp. 2–12.)

While Nidumolu, Prahalad & Rangaswami's research has focused on major non-tourism corporations, lessons learned are applicable to value chains and supporting businesses utilized more broadly in tourism (e.g., food services, transportation). And, even though the authors provide a framework to understand the stages businesses experience on a sustainability journey, which does appear critical, other researchers have explored the elements of "sustainable" business model conceptual framework. Critically evaluating what a sustainable business model (SBM) actually entails, compared to other models, may assist our research actions and provide transferable insights and applicability to tourism industries.

Sustainable business models

The dominant model of businesses has drawn on neoclassical economic theory which places shareholders and profit at the forefront of goals and operations (Brenner & Cochran, 1991; Key, 1999; Stormer, 2003). This is evident primarily in the wide adoption of business model research characterized by "economic and entrepreneurial theories of business innovations" (Upward & Jones, 2015, p. 100). One of the most widely accepted models is the business model canvas (BMC) (see Hanshaw & Osterwalder, 2015; Osterwalder & Pigneur, 2009), in which the "motivating logic" is profit (Upward & Jones, 2015). While several authors have called for the BMC to be extended to include sustainability requirements, it has been discounted "due to the inability to represent the complex social and ecological systems that are the context for all business in a meaningful way within the original Osterwalder ontology and canvas" (Upward & Jones, 2015, p. 100).

Stubbs & Cocklin's (2008) review of corporate businesses offered one of the first efforts to frame an SBM with six propositions incorporating sustainability:

- Organizational purpose that is defined in terms of ecological, social and economic outcomes;
- Measure performance and manifest this performance into a sustainability report, a social impact report and an environmental report, alongside financial indicators;
- Adopt a stakeholder view of the organization, rather than a shareholder view, with success linked to the stakeholders success, including local communities, suppliers, partners, employees and customers;
- Nature is treated as a stakeholder—negative impacts on it should be minimized or eliminated; promote environmental stewardship;
- Require visionary leaders who motivate cultural and structural change and a "sustainability mindset" throughout the organization;
- Require a systems and organizational-level perspective.

Their findings support concepts broadly accepted in current tourism sustainability literature, which consistently reinforce the critical inclusion of social, economic and environmental factors. However, as a case study, it provides limited comparability across different business sectors and industries. The Stubbs and Cocklin framework has been criticized because "important requirements of compatibility with natural and social sciences are omitted" (Upward & Jones, 2015, p. 101).

Boons and Lüdeke-Freund (2013, p. 13) progressed these ideas by addressing a component-based approach, using the four pillars of Osterwalder and Pigneur's (2009) BMC, and deriving a set of four normative requirements relevant to sustainability for consideration:

1. The value proposition provides measurable ecological and social value in concert with economic value;
2. The supply chain involves suppliers who take responsibility towards their own as well as the focal company's stakeholders;
3. The customer interface motivates customers to take responsibility for their consumption as well as for the focal company's ecological burdens to its customers;
4. The financial model reflects an appropriate distribution of economic costs and benefits among actors involved in the business model and accounts for the company's ecological and social impacts (Maas & Boons, 2010, in Boons & Lüdeke-Freund, 2013).

Identifying a framework for a strongly sustainable business model (SSBM), Upward and Jones (2015) established four "critical formative propositions" which they state are compatible with "fundamental and emerging knowledge in the introduced natural, social, economic, management, and psychological sciences" (p. 103). First, they describe the definition of a "strongly sustainable firm" as

> . . . an organization that only enabled strongly sustainable outcomes as one that creates positive environmental, social, and economic value throughout its value network, thereby sustaining the possibility that human and other life can flourish on this planet forever . . . Such a firm would not only do no harm, it would create social benefit while regenerating the environment ("doing good") to be financially viable ("doing well" . . .)." (p. 103)

Their definition implies that sustainability is congruent with the entire value network, which is part of major network initiatives such as the 10-Year Framework of Programmes on Sustainable Consumption and Production Patterns (10YFP) (see http://www.oneplanetnetwork.org/what-10yfp).

Their second critical proposition includes a definition of "value":

> Value is the perception by a human (or non-human) actor of a "fundamental need" . . . being met measured in aesthetic, psychological, physiological, utilitarian, and/or monetary terms.
>
> Value is created when needs are met via "satisfiers" . . . that align with the recipient's worldview and destroyed when previously met needs go unmet due to the withdrawal of satisfiers, the application of inappropriate ("pseudo") satisfiers, or the application of satisfiers that do not align with the recipient's worldview. (p. 105)

As a result, their findings suggest that an SSBM must "provide the organization a foundation for guiding the co-creation of value with all an organization's stakeholders: customers, shareholders, social, and environmental constituents and indeed all actors in the organization's value constellation . . ." (p. 105).

Their third critical proposition defines a business model as

> a description of how a business defines and achieves success over time, such that it provides a description of the logic for an organization's existence: who it does it for, to and with; what it does now and in the future; how, where and with what does it do it; and how it defines and measures its success. (p. 106)

The fourth critical proposition includes a definition of the concept of "tri-profit." Upward & Jones suggest the single tri-profit metric is the conceptual net sum of the "costs (harms) and revenues (benefits) arising as a result of a firm's activities in each of the environmental, social, and economic contexts in a given time period measured in units, appropriate to each" (p. 106). In addition, they go on to explain that "a tri-profit firm creates sufficient financial rewards, social benefits, and environmental regeneration, with sufficiency defined by stakeholders with the governance rights (power) to do so" (p. 106). These measures "would be expressed as accounting entries in a general ledger, for measures that use monetary units, and nonfinancial metrics, in various units of measure, or possibly unit-less . . . (p. 106).

In addition to these critical definitions, Upward & Jones further describe five *instrumental principles* that "any ontology of strongly sustainable business must fully conceptualize" (p. 106) while adhering to the four previously mentioned proposition definitions. These are summarized as:

Principle 1: An SSBM "must ensure that ethically and practically appropriate decisions . . . and actions . . . are described."

Upward & Jones determined that those that choose the "right" thing to do are those stakeholders "engaged with and by the organization in numerous ways" (p. 106). They found that to fully engage and ensure relationships between actors and their stakeholder roles, requires three related perspectives, including (1) the actors for whom the organization exists, or those that define the value the organization creates; (2) the actors who are directly affected by the organization (positively or negatively by value created or value destroyed); and (3) the actors involved in the ongoing processes the organization undertakes to create or destroy value. Governance covers such processes as supporting various actors impacted by the decision-making processes, including the "right" actions to take, a process for choosing the right actions, and actions are being done "right," must be described.

Principle 2: The boundaries of an SSBM must address:

- A social definition, based on agreement of the organization's purposes made by stakeholders who have, gain, or are granted power in the decision-making processes;
- A legal definition relative to ownership;
- Systems outside and within a business (including stakeholders, biophysical environment, human constructed social and monetary domains);
- The conceptual (knowledge), the social (relationships), and physical objects, which are those that need to be "owned" or "controlled"—the organization's capabilities and resources;
- The social (relationships), physical, and conceptual objects that are shared with other social constructs via the containing systems, described as formal agreements and realized in monetary flows with stakeholders (investments, payments, revenues, etc.), biophysical material flows to and from biophysical stocks and ecosystems services, as well as energy flows to and from the biosphere (for additional details, see Upward & Jones, 2015, p. 107).

Principle 3: Validation of an SSBM. An SSBM considers the requirements for sustainability of life, over as long a time period as feasible. The authors suggest long-term modeling scenarios (i.e., 30 years).

Principle 4: Upward & Jones acknowledge the necessary financial viability of a business model. This includes describing elements of financial viability and creating a record of both current and desired values (goals) (p. 108).

Principle 5: Upward & Jones suggest that the final principle contains social benefits and environmental regeneration. In addition to principles addressing stakeholders (described above), they identify the complexities associated with various stakeholders, such as different measures of success within different units, as well as clear definitions of who are and who are not legitimate stakeholders within individual humans, a collection of humans (firms, non-governmental organizations, governments, etc.), as well as non-human.

Upward & Jones have taken steps to move towards "improved ontology for models of successful strongly sustainable businesses" (p. 114), including inherent ecological, social and economic elements, demonstrating considerable complexities within each, which are often lacking in other models. They have also added the importance of the environmental, social and financial economy contexts, which support a systematic approach to creating an organization's values and extended linkages to sustainability.

Lüdeke-Freund & Dembek (2017) have suggested sustainable business model frameworks (SBM) be conceptualized as an emerging "integrated field" of research and practice. The integrated SBM conceptualization identifies slight overlap with other types of business models and fields, allowing for the provision of combining different tools from different disciplines, something researchers in sustainable tourism have been advocating for and have certainly highlighted within this text. In practice, they call for an identification of the necessary disciplines and skills to inform the SBM research program in their specific community of practice. Major beliefs and concepts can be represented by five features that would ensure effective contributions to sustainable development. These are:

- An explicit sustainability orientation integrating social, ecological, and economic concerns;
- An extended notion of value creation, questioning traditional definitions of value and success;
- An extended notion of value capture in terms of those for whom value is created;
- An explicit emphasis on the need to consider stakeholders and not just customers;
- An extended perspective on the wider system in which an SBM is embedded (Lüdeke-Freund & Dembek, 2017, p. 1670).

Sustainable business models and research

Researchers have identified several elements for the creation of a sustainable business model within the context of a complex systems approach. Many of these ideas can be applied to tourism businesses. In addition, the tourism community is beginning to coalesce around an understanding of critical aspects of sustainable development, which include the importance of entities within the tourism system (i.e., business, consumers, and destinations) working in tandem with frameworks addressing global sustainability problems (Budeanu et al., 2015). For example, if tourism businesses address their relevant impact on society through the lens of sustainable development goals—"thriving lives and livelihoods, sustainable food security, sustainable water security, universal clean energy, healthy and productive ecosystems, and governance for sustainable societies" (Griggs et al., 2013, p. 306)—the opportunities to evaluate the downstream impact of the business are high. However, it is clear that to address these goals, transdisciplinary efforts are necessary. Utilizing a systematic approach to measure and monitor businesses from inception to operations creates multiple opportunities for greater understanding.

Currently, one gap in understanding a sustainable business model appears to be integrating transdisciplinary collaboration in a meaningful way. Yet there are features prominent in the literature that ensure the possibility of "effective" contributions to sustainable development:

- The integration of social, ecological, and economic concerns (Joyce & Paquin, 2016; Lüdeke-Freund & Dembek, 2017; Upward & Jones, 2015);
- Extended notions of value creation and value capture, inclusive of measurable ecological and social value in concert with economic value (Boons & Lüdeke-Freund, 2013; Lüdeke-Freund & Dembek, 2017; Osterwalder & Pigneur, 2009);
- An explicit emphasis on the need to address multiple stakeholders (Joyce & Paquin, 2016; Lüdeke-Freund & Dembek, 2017; Upward & Jones, 2015);
- An extended perspective on the system in which a sustainable business model is embedded (Lüdeke-Freund & Dembek, 2017);
- Ideas on supply chains that involve suppliers who take responsibility for their own as well as the focal company's stakeholders (Boons & Lüdeke-Freund, 2013);
- Consideration of the customer–business interaction which motivates customers to take responsibility for their consumption as well as for the business's ecological burdens to its customers (Boons & Lüdeke-Freund, 2013);
- Financial models reflect an appropriate distribution of economic costs and benefits among actors involved and account for the company's ecological and social impacts (Maas & Boons, 2010).

Budeanu et al. (2015) suggest that while there is an abundance of research on tourism and sustainability, there is a "lack of criticality and application to practice" (p. 285), reaffirming the importance of interdisciplinary research coupled with wider social transformations (Miller & Twining-Ward, 2005), including a need for creativity in sustainability models that are adaptable to the new challenges of the present day. This could include the ability to incorporate the use of guiding frameworks, monitoring impacts of the frameworks, and critical evaluation of the implementation (e.g., Global Sustainable Tourism Council Industry Criteria: GSTC, 2015). Others also suggest the importance of applying tools such as life-cycle assessments (Pascual-González et al., 2016; Sonneman et al., 2015) to break boundaries of what has been typically identified as a single institution (Budeanu et al., 2015). Sun, Wu, and Yang (2018) suggest businesses are best viewed as an "individual organism whose purpose is to thrive, grow, and develop within its natural, built, and social environment" (p. 66). Yet they also see future research addressing these "self-sustaining organizations" as part of a greater network, such as destinations, to create "sustainable business ecosystems" more broadly (p. 70). They believe from an ecosystem perspective, the "adaptive capacity" of an ecosystems approach creates a situation whereby the business is more resilient to disruption overall. However, by strengthening a single business, as an element of a system, the resiliency of a system as a whole, theoretically, would be strengthened as well. These ideas are consistent with what is happening in the sustainable tourism business community. While many certification programs have been established to "certify the tourism business", there is a significant trend toward a broader effort at the destination level, identifying criteria that assist communities, small islands, cities, protected areas, and so on, with a framework for destination sustainability (see Bushell & Bricker, 2016; GSTC-D, 2018).

Tourism researchers would do well to incorporate critical measurement of value in tourism in non-economic terms, perhaps based on achieving, for example, the sustainable development goals that represent a global strategy to address societal and environmental challenges faced in the present day (see UNSDGS, 2018). Therefore, several ideas must be addressed to expand our understanding of a sustainable *tourism*-business model. From a research perspective, responsible consumption and production would address questions that delve into issues and questions, such as:

- The need to understand an individual business value chain and value network analysis to comprehend where and how dollars are actually spent;
- To answer what a sustainable value network might look like for tourism, with all its complexities and industries. How would such an analysis be attempted?
- To demonstrate what a business model would look like if it was based on building sustainable value networks;
- To define the role of a business in enhancing resilience of an SES as well as the role of a healthy SES in the resilience of an individual tourism business.
- How are feedback and interactions between businesses that are part of an SES understood, and how might a sustainable business model be developed as part of an SES?

In an effort to move forward, sustainable tourism business models must not only be aware of and responsible for the context in which they operate, but they must also understand their functioning as an organism within a larger system. It is incumbent upon sustainable business models to encompass an internal capability and outward responsibility. However, in the end, a business can only be sustainable when the context in which it operates is also sustainable (inclusive of the ecological, socio-economic, and cultural (i.e., attitudes towards consumption)) (Devereaux Jennings & Zandbergen, 1995). Hence, as Sun, Wu & Yang (2018) suggest: "If we use something from the system that is irreplaceable, like materials we use to provide energy for growth, we begin to place our lives—and the life of our business system that operates on the same principles at serious risk of destruction" (p. 69). As Leiper (2008) and others have reiterated, tourism and its complex, multifaceted set of industries is challenged with a "one size fits all" approach. The sustainability journey, the internal context of the operation itself, and the socio-ecological system it operates within, all summon the future of research and the promise of effecting positive change.

References

Berno, T. & Bricker, K. S. (2001). "Sustainable tourism development: the long road from theory to practice." *International Journal of Economic Development*, **3**(3): 1–18.

Boons, F. & Lüdeke-Freund, F. (2013). "Business models for sustainable innovation: state-of-the-art and steps towards a research agenda." *Journal of Cleaner Production*, **45**: 9–19.

Brenner, S. N. & Cochran, P. (1991). "The stakeholder theory of the firm: implications for business and society theory and research." Paper presented at the annual meeting of the International Association for Business and Society, Sundance, UT.

Bricker, K. (2018). "Positioning sustainable tourism: humble placement of a complex enterprise." *Journal of Park and Recreation Administration*, **36**(1): 208–14.

Budeanu, A., Miller, G., Moscardo, G. & Ooi, C.-S. (2015). "Sustainable tourism, progress, challenges and opportunities: introduction to this special volume." *Journal of Cleaner Production*, **111**: 285–94.

Bushell, R. & Bricker, K. S. (2016). "Tourism in protected areas: developing meaningful standards." *Tourism and Hospitality Research*, Special Issue *Tourism and Protected Areas: A Review of the Last Decade*, March 16, 2016, pp. 1–15.

Devereaux Jennings, P. & Zandbergen, P. A. (1995). "Ecologically sustainable organizations: an institutional approach." *The Academy of Management Review*, **20**(4): 1015–52.

Esty, D. C. & Winston, A. S. (2006). *Green to Gold: How Smart Companies Use Environmental Strategy to Innovate, Create Value, and Build Competitive Advantage*. New Haven, CT, Yale University Press.

Griggs, D., Stafford-Smith, M., Gaffney, O., Rockström, J., Ohman, M. C., Shyamsundar, P., Steffen, W., Glaser, G., Kanie, N. & Noble, I. (2013). "Sustainable development goals for people." *Nature*, **495**(7441): 305–7.

GSTC (Global Sustainable Tourism Council) (2015). "Global Sustainable Tourism Council criteria for destinations." Available at https://www.gstcouncil.org/en/gstc-criteria/criteria-for-destinations.html (accessed August 2, 2018).

GSTC-D (Global Sustainable Tourism Council) (2018). "Global Sustainable Tourism Council criteria for destinations." Available at https://www.gstcouncil.org/en/gstc-criteria/criteria-for-destinations.html (accessed August 2, 2018).

Hanshaw, N. & Osterwalder, A. (2015). "The business model canvas: why and how organizations around the world adopt it. A field report." Strategyzer. Available at https://s3.amazonaws.com/strategyzr/assets/research_report.pdf.

Higham, J., Cohen, S. A., Cavaliere, C. T., Reis, A. & Finkler, W. (2016). "Climate change, tourist air travel and radical emissions reduction." *Journal of Cleaner Production*, **111**: 336–47.

Joyce, A. & Paquin, R. L. (2016). "The triple-layered business model canvas: a tool to design more sustainable business models." *Journal of Cleaner Production*, **135**: 147–86.

Key, S. (1999). "Toward a new theory of the firm: a critique of stakeholder 'theory.'" *Management Decision*, **37**(4): 317–28.

Law, A., De Lacy, T., Lipman, G. & Jiang, M. (2016). "Transitioning to a green economy: the case of tourism in Bali, Indonesia." *Journal of Cleaner Production*, **111**: 295–305.

Leiper, N. (2004). *Tourism Management* (3rd edn). Sydney: Pearson Education.

Leiper, N. (2008). "Why 'the tourism industry' is misleading as a generic expression: the case for the plural variation, 'tourism industries.'" *Tourism Management*, **29**(2): 237–51.

Likorish, L. J. & Jenkins, C. L. (1997). *An Introduction to Tourism*. Oxford: Butterworth-Heinemann.

Lüdeke-Freund, F. & Dembek, K. (2017). "Sustainable business model research and practice: emerging field or passing fancy?" *Journal of Cleaner Production*, **168**: 1668–78.

McLennan, C. J., Becken, S. & Watt, M. (2016). "Learning through a cluster approach: lessons from the implementation of six Australian tourism business sustainability programs." *Journal of Cleaner Production*, **111**: 348–57.

Maas, K. E. H. & Boons, F. A. A. (2010). "CSR as a strategic activity: value creation, redistribution and integration." In C. Louche, S. Idowu & W. Leal Filho (eds.), *Innovative CSR: From Risk Management to Value Creation*, pp. 154–72. London: Greenleaf.

Miller, G. & Twining–Ward, L. (2005). "Tourism optimization management model." In G. Miller & L. Twining-Ward (eds.), *Monitoring for a Sustainable Tourism Transition*, pp. 201–32. Wallingford: CABI.

Milne, S. (1998). "Tourism and sustainable development: exploring the global-local nexus." In C. M. Hall & A. A. Lew (eds.), *Sustainable Tourism: A Geographical Perspective*, pp. 35–48. Harlow: Longman.

Nidumolu, R., Prahalad, C. K. & Rangaswami, M. R. (2009). "Why sustainability is now the key driver of innovation." *Harvard Business Review*, **87**(9): 1–12.

O'Fallon, C. E. (1994). "The role of central government in tourism: a public choice perspective." PhD thesis: Lincoln University.

One Planet Network (2018). Available at http://www.oneplanetnetwork.org/sustainable-tourism/about (accessed July 1, 2018).

Osterwalder, A. & Pigneur, Y. (2009). *Business Model Generation: A Handbook for Visionaries, Game Changers, and Challengers*, ed. T. J. Clark & A. Smith. Hoboken, NJ: Wiley.

Pascual-González, J., Guillén-Gosálbez, G., Mateo-Sanz, J. M. & Jiménez-Esteller, L. (2016). "Statistical analysis of the ecoinvent database to uncover relationships between life cycle impact assessment metrics." *Journal of Cleaner Production*, **112**: 359–68. https://doi.org/10.1016/j. jclepro.2015.05.129.

Sonneman, G., Gemechu, E. D., Remmen, A., Frydendal, J. & Jensen, A. A. (2015). "Life-cycle management: implementing sustainability in business practice." In G. Sonnemann & M. Margni (eds.), *Life Cycle Management, LCA Compendium – The Complete World of Life Cycle Assessment*, pp. 7–22. Dordrecht: Springer.

Stormer, F. (2003). "Making the shift: moving from 'ethics pays' to an inter-systems model of business." *Journal of Business Ethics*, **44**: 279.

Stubbs, W. & Cocklin, C. (2008). "Conceptualizing a 'sustainability business model.'" *Organization & Environment*, **21**(2): 103–27.

Sun, J., Wu, S. & Yang, K. (2018). "An ecosystemic framework for business sustainability." *Business Horizons*, **61**: 59–72.

Swarbrooke, J. (1999). *Sustainable Tourism Management*. Wallingford: CABI.

UNSDGS (United Nations Sustainable Development Goals) (2018). "Sustainable development knowledge platform." Available at https://sustainabledevelopment.un.org/sdgs (accessed October 11, 2018).

Upward, A. & Jones, P. (2015). "An ontology for strongly sustainable business models: defining an enterprise framework compatible with natural and social science." *Organization and Environment*, **29**(1): 97–123.

10 The human health dimensions of sustainable tourism

Monika M. Derrien, Lee K. Cerveny and Kathleen L. Wolf

Introduction

Over the millennia restorative places have lured people, many of whom have traveled to destinations that promised to reinvigorate their health. From the mid-nineteenth to early twentieth century, many people with tuberculosis traveled to sanatoriums in the countryside, hopeful for the curative powers of fresh air and nature. These facilities—often located in places such as the Swiss Alps and New York's Adirondack Mountains—did aid in the recovery from mild cases of tuberculosis, but the development of antibiotics in the 1940s eventually rendered them obsolete (Frith 2014). However, the belief that natural elements and environments can promote health has endured. In addition to health-promotion travel, some tourists travel specifically for clinical treatments at medical facilities far away from home, or seek emergency medical care during their travels. Others derive health benefits as secondary effects, such as stress reduction during a beach vacation. This chapter explores the many ways tourism and health are inextricably linked.

From ancient to modern times, salutogenic tourism networks have never occurred in a vacuum. Destinations are developed and marketed by community and industry boosters, relying on more than a traveler's chance encounter with the proverbial fountain of youth. Local businesses facilitate the movement and accommodation of travelers and collectively craft opportunities that range from treatment and therapy opportunities to facilities for recreation, relaxation, or fun in the sun. As a result, travelers' experiences of personal health and well-being occur in many types of places, including urban parks, community healthcare facilities, the rural countryside, beaches, cultural heritage sites, and pedestrian greenways.

This chapter proposes health concepts and considerations that tourism planners and researchers can use to help build resilience into tourism systems. It explores the health outcomes of intertwined social, economic and ecological systems for visitors, residents and workers. We first provide background and definitions for tourism and health, followed by a conceptual framework of related benefits. Finally, we use the framework to propose research opportunities that can better integrate human health goals into tourism systems across a variety of wildland, rural and urban settings.

Background

Tourism and its human, natural, and built capital

Tourism includes travel for leisure, work and other purposes, and travel motivations may include the pursuit of health outcomes as primary or secondary purposes. For example, someone might choose to travel to Japan for a professional conference, in part because they are intrigued by the opportunity to visit one of the country's official forest bathing sites or temples (Li 2018). Someone traveling to Thailand for an inexpensive root canal procedure may spend some of their savings by extending their travels to include time on a tropical island (Moon 2017). Many different types of travel, therefore, even those that are oriented to clinical or therapeutic purposes (and less so to leisure), are still considered tourism (United Nations World Tourism Organization 2010). Health benefits are often gained as positive externalities of tourism, whether or not those benefits are explicitly appreciated.

The term "vacation" is central to conceptions of tourism. Rooted in the Latin "vacare," meaning "to be unoccupied," the notion of a vacation often evokes thoughts of free, unstructured time to be filled with travel experiences. When someone says, "I'm going on vacation," most assume they are traveling somewhere. Such free time may be seen as a "vacuum" to fill with enjoyable and restorative experiences, or simply a temporary "vacating" of their usual residence.

The neologism "staycation" emerged in the early 2000s to describe time taken off from work expressly to enjoy close-to-home vacation-like experiences while avoiding their greater time and monetary costs (White 2011). Being a "tourist in one's own city" generally refers to experiencing museums, parks and other attractions that residents may not routinely visit. While this chapter's definition of tourism suggests displacement, the "staycation" notion encourages thinking about how the benefits of tourism can be experienced more sustainably and equitably by a broader swath of the population.

The following sections briefly describe the human, natural and built capital of tourism systems and communities.

Human capital

The human capital of tourism systems is composed of individuals and groups engaged in economic and social relationships that enable the movement of travelers. Tourists themselves are generally the most visible face of the tourism system. Their demands, spending, activities and presence shape landscapes and communities in many ways. Tourism providers are another component of the human capital of tourism. Tour guides, travel agents, retailers, hoteliers and others earn income directly from the tourist trade. Other workers play a support role, producing food for tourist consumption, or manufacturing local products for purchase. Trade organizations, such as chambers of commerce or convention and visitors bureaus, often are critical in bolstering local tourist industries. Local, state and federal government officials formulate policies about land use regulations and zoning, permitting and siting new development, and taxes. National, state and provincial governments invest in tourism departments that market their regions, and influence tourism through border regulations and travel restrictions, managing commerce through tariffs and taxes, and regulating transportation and accommodations. Public land managers implement policies governing recreation and cultural heritage sites, influencing the diversity of visitors through the facilities, services and access that is available. Non-governmental organizations (NGOs) may contribute to tourism by providing political support, advocacy and programming, and may serve as a watchdog monitoring tourism's effects on the natural environment or indigenous communities (Simpson 2008). Finally, healthcare providers serving tourism communities also play a role in the system. Hospitals may market specialized procedures to attract patients from outside of their base community. These providers also respond to diseases or injuries experienced by travelers.

Tourism development has implications for the communities in which it occurs, providing opportunities and benefits for residents, but also leading to a variety of changes associated with visitor volume, concentration, duration and frequency; commoditization of local traditions; and other socio-cultural factors that have been well documented (Smith 2012; see also this volume, Cerveny & Miller, Chapter 11). Especially in small rural communities, a quickly expanding tourism sector can be a conspicuous presence (Park et al. 2018). Whether or not they are engaged directly in tourism, residents' health may be affected. Tourism development can interfere with their local opportunities to harvest foods for subsistence or livelihood, to pursue traditional forms of outdoor recreation, or to spend time in favorite outdoor spaces. At the same time, tourist development can provide new opportunities for residents to take advantage of facilities designed for visitors, but available to residents (Cerveny 2008). Municipal officials may negotiate directly with global tourism providers to provide infrastructure in exchange for the promise of tourists (Ruhanen 2013), but when power relationships are unequal, local communities may find themselves at a disadvantage (Stonich 1998).

The distribution of tourism experiences across the world's population is an important consideration. Many who work in the tourism sector may never travel themselves: many people in the world never have the opportunity to travel beyond the bounds of their village, city or country (Lee & Jamal 2008). Many who live adjacent to tourism destinations may not partake in the activities offered to wealthier visitors. Who is allowed or chooses to travel (and with whom), where they go, and what they do during their travels are matters of financial ability as well as cultural and social norms and national policies.

Natural capital

The natural capital of the tourism system comprises the environmental features and phenomena that attract and sustain visitation—from leafy boulevards that make cities enticing to explore, to "bucket list" destinations in wilder places, such as the Great Barrier Reef. Even in places where natural capital is not featured or appreciated, local ecosystems sustain life-supporting materials and processes. While this chapter considers tourism broadly, it takes a particular interest in tourism places that have been developed around natural features and environments, because of the growing body of research that demonstrates the beneficial health outcomes from experiences of nature (Frumkin et al. 2017; Wolf & Robbins 2015).

Built capital

Even for destinations offering nature-based experiences, tourism requires significant built capital, from transportation networks, to lodging and dining establishments, to parking lots and restrooms. Places with inadequate or unsafe infrastructure endanger the health of visitors during their travels and potentially afterwards, with similar risks for residents. More positively, the built environment, including art, design, housing styles and civic spaces is an important part of the cultural context of tourism. Many tourism places that "get on the map" because of extraordinary natural features evolve and expand services, some having little to do with the natural attractions, such as theme parks or outlet malls. Other communities intentionally construct their built environment to reflect natural and community heritage, such as through museums or interpretive trails. The costs of planning for and developing infrastructure in support of tourism often fall on local and regional governments, which must weigh these investments alongside other social programs that primarily serve residents. Infrastructure and facilities designed to support tourism also can lead to other investments that provide jobs or serve local needs. For example, a road built to access a waterfall and eco-tourist lodge in West Africa may also be used by villagers traveling to market, or palm oil harvesters accessing their crops.

The social determinants of health

Health is defined as "a state of complete physical, mental and social well-being and not merely the absence of disease or infirmity" (World Health Organization 2016, p. 1). Health providers and researchers are increasingly focused on the social determinants of health as a way to understand the population-level factors that influence health outcomes. The social determinants of health are "the conditions in which people are born, grow, live, work and age" (World Health Organization 2015, p. 4), and are considered "responsible for a major part of health inequities between and within countries" (Commission on Social Determinants of Health 2008, p. 1). The five main social determinants of health proposed by the US Department of Health and Human Services (2018) are: (a) economic stability; (b) education; (c) social and community context; (d) health and healthcare; and (e) neighborhood and built environment.

The built, natural, social, intellectual and economic environments of everyday life—including nearby nature and spaces that provide opportunities for social engagement—are the "constants" that can promote good health. Tourism typically offers episodic experiences (although often memorable and impactful), but can create enduring changes in the lives of residents and tourism workers. Local conditions, including physical and social infrastructure and institutions, are highly influenced by the industry—from the jobs tourism supports, to the availability of healthcare, to access to public lands. Recognizing this, the United Nations World Tourism Organization works to advance tourism's contribution to poverty reduction and development, in pursuit of its Sustainable Development Goals for 2030 (United Nations 2017).

Overarching questions

Tourism systems are dependent on healthy people to provide the human capital that sustains them. Integrated considerations of the health and quality of life of both visitors and residents of tourism destinations are important, but the academic literature on tourism's effects on the quality of life for these two populations has not been well integrated (Uysal, Perdue & Sirgy 2012). Do tourism networks allow for greater access to health and wellness resources for all? How is the health and well-being of residents shaped by different types and degrees of tourism development? What are the health benefits tourists derive from their travel experiences? Under what circumstances does tourism contribute to greener, safer or more pedestrian-oriented cities? Are the employment opportunities generated by tourism contributing to healthy communities? The conceptual framework proposed in this chapter offers structure for thinking about these questions.

Conceptual framework of tourism–health interactions

Our conceptual framework proposes the many ways that human health intersects with tourism systems, to help guide future research and thinking about how health is being assessed in relation to tourism as a sustainable development strategy. The framework describes six types of health–tourism interactions, and linkages to social determinants of health (Table 10.1).

1 "Nearby" and "destination" nature

People have long traveled to destinations that offer restorative natural features such as mineral springs or the seaside. Contemporary health and wellness tourist networks market destinations that offer experiences in nature, such as *shinrin yoku*, the practice of forest bathing originating in Japan, and wellness retreats offering activities such as outdoor yoga and hiking. Clinical therapeutic programs, such as wilderness therapy for teens with mental health or behavioral struggles, and outdoor-activity-based programs for veterans with post-traumatic stress disorders, share some similarities with health and wellness tourism, but typically focus on more formal treatment rather than prevention.

Table 10.1 Conceptual framework of tourism–health interactions

Tourism–health interaction	Description	Social determinants of health
(1) "Nearby" and "destination" nature	Tourism systems can connect travelers with the restorative benefits of nature contact in both health-promotion and treatment contexts. Residents may benefit from investments in nearby parks and urban greenery through tourism development.	Neighborhood and built environment. Social and community context.
(2) Built and community environment	Opportunities for physical activity and social interaction are shaped by the built environment; tourists' mobility and activities are related to neighborhood walkability and the design of public spaces. The built environment can be transformed by tourism development; effects on residents vary, including altered senses of place and local culture.	Neighborhood and built environment. Social and community context.
(3) Mental health and function	Tourism provides travelers with opportunities to engage in new environments, reducing stress, restoring attention and promoting cross-cultural learning. Travelers build cohesion through spending time with family and friends, and interacting with other tourists or residents. Tourism workers and residents may experience higher stress levels during peak tourist seasons.	Education. Social and community context. Neighborhood and built environment.
(4) Health risks and access to healthcare	New environments for travelers can result in unfamiliar health risks. Treatment-oriented tourism can improve access to procedures or medications. Tourism can also create new funding channels and access to healthcare.	Health and healthcare.
(5) Tourism-related employment	The tourism industry provides employment opportunities, which, although variable in quality, can increase income and improve access to healthcare for some residents.	Economic stability. Health and healthcare.
(6) Ecosystem health and environmental sustainability	Tourism can be a motivator for conservation, but also can degrade local environments and contribute to greenhouse gas emissions.	Social and community context. Economic stability.

Both clinical/therapeutic and preventive/restorative nature contact have demonstrated benefits across the landscape gradient, from nearby nature in urban settings to remote, wild lands (Twohig-Bennett & Jones 2018). Multiple disciplines are interested in the benefits of nature contact, developing increasingly sophisticated methods of measuring its outcomes. In a recent literature review, Frumkin et al. (2017) identified studies where general population measures showed improvements in well-being and health across countries or communities. The studies reviewed provide evidence of specific outcomes possible during nature-based tourism, such as lower blood pressure, better sleep, improved heart health, improved immune function, and other positive outcomes that span the human life cycle. Social and mental health benefits included reduced stress, greater happiness, increased life satisfaction, and increased social connectedness with fellow travelers.

As the world's population and land base becomes more urbanized (United Nations 2018), there is an increasing interest in creating spaces for the therapeutic and restorative power of nature contact in cities. Since much tourism is centered in cities, a resident's "nearby nature" can become a shared amenity for a tourist. The nature of cities—including green infrastructure, street trees and canopy cover, water bodies, greenways, and public parks and gardens—is a social determinant associated with a wide range of positive health effects, including crime deterrence (Kondo et al. 2017). Proximity to green spaces is associated with increased physical activity (Giles-Corti et al. 2005), and activity in natural settings provides more benefit compared to indoor exercise (Coon et al. 2011). Residents living in nature-rich environments may derive ongoing effects from regular nature contact, while the short-term exposure of visitors may generate different outcomes.

Questions for future research

1.1 **Dose and duration**: What is the dosage needed to initiate and sustain health benefits, including the type of environment, frequency and duration of experience, and unique needs of specific populations? How long do the benefits of nature exposures of different types, intensities, and durations last? What is the consequence of the short-term "peak" experiences typical of tourism, compared to the presence of protective elements of everyday life (e.g., close-to-home gardens, trees, and water bodies)? What are the causal pathways of nature benefits, and how can public land managers enhance their provision? Tourism planners and researchers might study the comparative effects of nature contact on residents (regular, but perhaps brief), tourists (episodic, and longer in duration), and nature-based tourism operators (regular, and longer in duration).

1.2 **Human–nature connections across the landscape gradient**: How can tourism be planned to contribute to connections between people and nature that occur both at tourism destinations and at closer-to-home places? Beatley (2017) describes the "nature pyramid," with destination-tourism experiences of nature at the top of the pyramid, and everyday, neighborhood-scale experiences forming the base of a healthy "nature diet." How can tourism planners and health professionals help shift perceptions of nature being "out there" and nurture appreciation for and bonds with nature that is closer to home?

1.3 **Designing nature-based spaces**: Understanding health–nature connections across the gradient can inform the development of better-built capital for tourism. What are the essential nature-based spaces, facilities and programs that can encourage multiple benefits for visitors and local populations?

1.4 **Travel-based clinical therapeutic programs**: The benefit of nature contact for people in therapeutic programs is likely supplemented by other therapies before, during and after the nature-centered experience. Can the effects of nature contact be disentangled from other therapeutic factors and benefits pathways, such as group bonding and activity engagement? How can such knowledge be used to refine programming? How can public land managers better support such programs?

2. Built and community environment

Many persistent human health problems—diabetes, cardiovascular disease, obesity—are related to increasingly sedentary lifestyles promoted by the built environment in indoor and outdoor settings (de Rezende et al. 2014). Other concerns—depression, stress, and a host of other physical manifestations—are linked to social isolation and loneliness (Maas et al. 2009). Built environments at all scales can encourage beneficial physical and social activity: bicycle and pedestrian infrastructure promote active transport; waterways allow travel by non-motorized watercraft (or skates in winter); public spaces encourage informal social interactions (Peters, Elands & Buijs 2010); and interpretive materials entice people to explore by foot. Some tourism destinations encourage walking simply because their pre-vehicular layouts are a comfortable scale for pedestrians.

Urban planners, landscape architects, public land managers and businesses all contribute to creating destinations conducive to active transport, movement-oriented leisure activity and social interaction. Many tourist activities center on physical activity, from walking tours of historic sites to snorkeling on coral reefs. Local businesses may offer equipment rentals or guided trips. Some protected areas only allow non-motorized activities, in effect encouraging active modes of travel. But active tourism can present challenges to land managers and residents. For example, during destination events such as trail races, parking lots, trails and facilities often become overwhelmed by the large crowds.

Community-level social factors—including support systems, community engagement, safety, and social integration—can be affected by tourism. Residents may experience disagreements about tourism development and altered community dynamics. They may be challenged by increases in transportation and parking congestion; undesirable activities such as crime, public drunkenness, drug use, gambling and prostitution; and conflicts from differences in social and moral values

(Deery, Jago & Fredline 2012). Residents may visit special places less often because of crowding at popular sites.

Positive effects to the built and social environment from tourism development may include increased shopping, entertainment and recreational facilities; increased opportunities to socialize with locals and tourists; and increased sense of pride in community (Deery, Jago & Fredline 2012). If new developments offer open access, residents may share in the benefits of improved parks, recreation and transportation infrastructure, but real and perceived barriers may also limit access across racial, ethnic and socioeconomic populations (Giles-Corti & Donovan 2002).

Questions for future research

2.1 **Integrated tourism planning**: What tourism strategies and development can best promote desirable health outcomes and discourage undesirable ones? What are tourism's effects on ecosystem health, the built environment, and downstream socioeconomic effects on residents and communities? Can health impact assessments be better incorporated into integrated tourism planning processes, to ensure that health is an explicit consideration (Bauer 2008)?

2.2 **Physical activity promotion within the built environment**: How can best practices that promote physical activity be better integrated into sustainable tourism planning (e.g., building design, commercial district design, active transportation infrastructure)? How can these processes ensure that benefits are also experienced by residents?

2.3 **Destination choices**: How do considerations of health influence travel decisions about where to go, where to stay and what to do? Does having infrastructure and programming for physical activities factor into tourists' decisions, from macro choices (does a destination market opportunities for active recreation) to micro site choices (is a lodging option located in a walkable and safe area)?

3. Mental health and function

Nature experiences provide benefits of increased mental health and function, including alleviating mental fatigue, restoring attention, reducing stress, improving affect, increasing social cohesion and creating learning opportunities (Bratman et al. 2012). Social interactions play a central role in the beneficial relationship between green space and health, particularly in reducing feelings of loneliness (Maas et al. 2009). Tourists that spend time in nature-rich destinations can benefit from these outcomes, including the potential to form meaningful attachments to places.

Some countries and employers send stressed workers to spa destinations for restoration and recuperation, institutionalizing a connection between health and tourism that recognizes the mutual benefit for employers and their workers (Naraindas & Bastos 2011). Trips to fishing lodges, yoga retreats, and surf camps are sometimes offered by corporations as employee rewards and incentives. Corporate retreats, which combine work and recuperation, are often located in settings rich in natural amenities.

People report feeling happier, healthier and more relaxed in the days and weeks after they return from pleasure travel, and increased social cohesion and bonding among friends and family in a travel context can deliver health benefits (Chen & Petrick 2013). If people disconnect from their work responsibilities during travel, they may experience greater cognitive benefits and reduced stress (Fritz and Sonnentag 2006). Yet forfeiting vacation time is a recent trend (NPR 2016), and negative health effects are substantial; in Japan, the prevalence of *karoshi*—death from overwork—has increased (Li 2018).

Travel affords the opportunity for novel, self-directed activities, learning, gaining new perspectives, and other mentally stimulating and challenging experiences. Even travel that does not include a complete escape from responsibilities offers cognitive benefit; for example, students who have studied abroad show more creative thinking than those who have not (Lee, Therriault & Linderholm 2012). Tourism experiences centered on volunteer activities can forge new relationships and social capital, as well as allow for the development of new skills and relationships to place (Coghlan & Gooch 2011). A growing tourism sector, building on "the experience economy," offers travel experiences in immersive settings to learn new activities. During different phases of the life course, travel may serve considerably different purposes and desires: parents traveling with young children may seek different activities than students studying abroad or retired seniors. Some cultures encourage "rites of passage" travel experiences for young adults, such as through Birthright Israel, Mormon missions, and the Australian aboriginal walkabout. Programs such as Road Scholar offer seniors immersive learning experiences through travel.

When they have opportunities for interaction, residents and tourists may learn from each other and gain new cultural perspectives. Yet, local cultures risk becoming "canned," as some residents experience pressure to present an abstract or stereotypical version of their culture for tourist consumption (Urry 1990). The appropriation of local traditions and commoditization of indigenous or local identities by tourism proponents can contribute to low self-esteem. This can have a negative effect on identities and reinforce power structures that are unfavorable to local communities (Morgan & Pritchard 1998).

Mental health considerations are important for residents and workers. Stressors may be introduced into residents' everyday lives as community character changes, and perceptions of crowding may increase with tourist activity (Deery, Jago & Fredline 2012). Seasonal fluctuations of visitors can be particularly stressful, as workers may be overwhelmed during the peak season and feel uncertain about their economic well-being during the off-season, particularly if alternative economic opportunities are scarce. In destinations that attract visitors from other countries, residents may experience stress from the presence of strangers or being surrounded by people whose language, customs, appearances and practices are very different from their own (Smith 2012).

Questions for future research

3.1 **Enduring cognitive benefits**: What are the lasting mental health and cognitive benefits for tourists, considering tourism experiences of different types, durations and activities? Considering the environmental impacts of the transportation sector, are there ways to promote those benefits in closer-to-home experiences (or as part of a "staycation")?

3.2 **Place relationships and interpretation**: While built environments, such as transportation networks, enable tourism experiences, it is often the natural environment that draws people to visit (grand waterfalls, biodiversity in the lush rain forest). Can environmental interpretation help foster deeper relationships between tourists and the natural environments that attract them? Do connections with different types of places result in varied mental health benefits?

3.3 **Enduring activity and participation**: Do people who learn new activities while on vacation (such as yoga or a new language) continue to participate in it when they return home? Is the activity sustained at the level needed to provide positive health effects?

3.4 **Resident well-being**: How can tourism's stressors for local residents and workers be alleviated through planning that focuses on spatial or temporal solutions for tourist interactions and concentrations?

4. Health risks and access to healthcare

The adequacy of a community's healthcare infrastructure for emergency care is an important consideration in tourism planning (Richter 2003). Travelers face a range of health risks, especially international travelers who may be unaccustomed or ill-adapted to the climate, altitude, microbes, and hygiene and sanitation practices in the countries they visit (Walker, LaRocque & Sotir 2017). Because of their mobility, travelers may also transport and transmit infectious diseases to local populations (Richter 2003).

Emergency healthcare is especially relevant in small towns and cities that serve as ports for cruise ships that receive large populations of older visitors—a demographic that has a higher likelihood of requiring emergency medical services. Tourists often engage in more risky behaviors while on vacation, which might make them susceptible to injury or illness (Pizam, Reichel & Uriely 2001). Outdoor activities such as whitewater kayaking and mountain biking can result in backcountry injuries requiring emergency medical response and evacuation (Flores, Haileyesus & Greenspan 2008). The businesses and organizations that outfit or guide these activities, as well as adjacent communities, may develop liability and risk plans to anticipate emergencies, engage employees in understanding risk, and set up response protocols.

Wildfires and other natural disasters can also impair human health (e.g., respiratory problems from wildfire smoke) and cause disruption to international and national travel circuits. The increase in frequency and severity of extreme weather events due to climate change can be expected to increase travel disruptions and health impacts (Koetse & Rietveld 2009). Preparedness can help, with health and tourism agencies, as well as chambers of commerce, playing important planning and communications roles.

Taxes on tourism revenues have the potential to contribute to community health by funding social service programs, health clinics and education, although some communities struggle with ensuring that tourism revenues actually fund the programs that politicians and community leaders promised when advocating for new development (e.g., Chason 2018). In Alaska, per-passenger taxes from the cruise ship industry are used for substantial and durable improvements in local communities, but there are concerns about whether residents actually benefit from them. In their study of casino development in tribal communities, Kodish et al. (2016) found that health outcomes have been mixed despite increased household incomes and community social service and wellness programs. In developing regions such as Africa, tourism industry promoters may negotiate for use of village resources by offering to build health clinics that serve residents. Such trade-offs are important areas of leverage for local leaders and require regular re-evaluation.

Tourism networks also provide opportunities for "medical tourism" or the practice of obtaining medical treatment outside of one's home country. Cosmetic surgery, dentistry and heart surgery are common procedures (Nguygen & Gaines 2017). Motivations include access to treatments not available locally, access to higher quality or less expensive treatments than are available locally, or a desire to undergo treatment closer to family or friends. The distinction made between medical tourism and health-promotion tourism is that medical tourism involves treatment for clinical or medical conditions, while health-promotion tourism involves seeking health-promoting environments and activities (Connell 2013), although some destinations or programs may achieve both (e.g., weight-loss camps that teach healthy lifestyle habits). Questions about who is traveling where, and what their motivations and health outcomes are, remain under-researched, with industry estimates often presenting inflated numbers (Connell 2013).

Questions for future research

4.1 **Information networks**: What information networks are needed to better communicate the risks of fires, droughts and storms to national and international travelers? In the age of "over information" from social media platforms and news outlets, and an increased occurrence and intensity of weather events, how can health and safety advisories better reach travelers? Are there ways for improved integration across the tourism system to provide information and coordination of travel alternatives before or during the occurrence of major weather events?

4.2 **Tourism on tribal lands**: How can casino development, especially on tribal lands, be implemented in more socially and culturally sustainable ways, thus reducing negative health effects? Can comparative case studies reveal the best practices to promote more sustainable outcomes?

4.3 **Understanding medical tourism**: Accurate estimates of the extent of medical tourism, the roles of tourism companies as intermediaries, and health outcomes are lacking (Connell 2013). What are the demographics, experiences, satisfaction and motivations of medical tourists? What is the impact of medical tourism on local communities' revenues and health infrastructure? What ethical issues exist? Does servicing tourists impede local access to health services, or does it allow for improved local service quality?

5. Tourism-related employment

Employment and working conditions are important factors shaping local economic stability, a social determinant of health. Job creation is often at the forefront of planning for tourism development, and in some cases, local government officials and industry promoters advance a "rhetoric of despair," calling for the need for new industry to revive "dying towns" (Stokowski 1996). In practice, however, tourism jobs often pay low wages compared to other industries (Lacher & Oh 2012). A tourism industry can provide a range of employment opportunities, depending on the location, the industry's seasonality and the level of workforce specialization required. Tourism jobs do provide part-time or seasonal jobs that suit some workers, especially students, care-takers (often women) and seasonal workers from other industries (Butler 2001). Tourism allows some workers in a patchwork economy to mix tourism employment with other jobs and income-producing activities (Vandegrift 2008). Tourism can impact local economies beyond employment-related economic stability, including the cost of living, redirected local government revenues and availability of services.

Some tourism-related activities also allow people to remain connected to their land and craft. Revenue from agritourism, for example, can supplement the income of small working farms, allowing continued operation and helping to stabilize seasonal and annual fluxes in revenues (Nickerson, Black & McCool 2001). Guest ranches, where people stay on site, ride horses and participate in ranch chores, enable hosts to share a lifestyle and continue to maintain their ranches.

Some jobs may promote employee health, such as the work of outdoor guides and activity instructors. People working in the ski industry, for example, may enjoy working outdoors and maintaining a connection with nature or a particular mountain. The lifestyle of these service-sector jobs often sustains tourism operations despite a lack of competitive remuneration. In addition, some volunteers work for short periods on organic farms in exchange for lodging and cross-cultural educational experiences (e.g., World Wide Opportunities on Organic Farms, or WWOOF), extending their travel budget. These experiences bridge work and leisure, creating opportunities for health-promoting activities within work and volunteer experiences.

Questions for future research

5.1 **Local economic impacts**: How can local communities be better prepared to retain greater local economic returns, versus distant firms and corporate shareholders? How can tourism dollars be better leveraged and partnerships developed to sustain long-term health-promoting infrastructure and programming in communities?

5.2 **Equity of economic benefits**: How can the distribution of the economic benefits of tourism be made more equitable, to improve remuneration, work conditions and opportunities for career advancement in service-sector jobs?

5.3 **Occupational exposures**: How can the beneficial occupational exposures of nature contact be best measured for outfitters, guides and activity instructors? What are the health benefits and challenges for "migratory" workers in the tourism industry, such as people who work seasonal jobs? How might nature contact and physical activity be understood and promoted to increase resilience for tourism's employment base?

6. Ecosystem health and environmental sustainability

The complex interactions of tourism, land conservation, environmental impacts and human health benefits generate many important questions for future research. Healthy, intact ecosystems have positive effects on human health in direct ways; for instance, they provide and regulate the quality of the water and air humans need to survive. Tourism, or the promise of tourism, can be a motivation and enabler for the protection of lands, both within public and private domains (Boley & Green 2016). Economic motivations for developing a nature-based tourism sector can lead to decisions that support ecological integrity, such as the preservation of land that otherwise would be mined, logged and/or converted into agricultural use.

Tourism is generally considered "lighter" on the land than other uses that land is being "protected" against. Yet such assessments are dependent on the scale and type of tourism activities being considered as well as those of the alternative land uses. Increased visitation can lead to the damage of natural and cultural features due to user pressures, human-caused wildfires, vandalism and private exploitation, lessening the quality of experience for others and damaging or destroying the very resources that motivated preservation in the first place. Tourism can cause general "wear and tear" on public lands and create issues with human waste, trash and erosion. Tourism development around natural areas can disturb wildlife habitat and populations (this volume, Cerveny & Miller, Chapter 11; Davenport & Davenport 2006).

Concerns extend to built environments. Tourism development may trigger amenity migration and residential development that results in changing land use priorities (Ooi, Laing & Mair 2015). Urban planners and land managers struggle with the challenges of peaks and flows of visitation and cross-cultural communications about desired practices and behaviors in public spaces. Influxes of visitors can create waste management concerns in communities whose sewer and solid waste disposal systems were not designed for visitor volumes, resulting in the pollution of rivers and shorelines.

Travel is a major contributor to greenhouse gas emissions. Even if tourism operations are committed to sustainability practices (e.g., green building design, reducing greenhouse gas emissions and water consumption, diverting waste from landfills), the transportation sectors that connect them to their customers generally are major contributors to climate change (Higham et al. 2016). The human health impacts of climate change are discussed at length elsewhere but have many negative dimensions, including increased vector-, food-, and water-borne diseases, air quality, flooding, and mortality and morbidity due to heat waves (Haines et al. 2006).

Questions for future research

6.1 **Sustainability of operations**: What are the long-term consequences of concentrated human visitation and use in different environments? Given the greenhouse gas emissions associated with travel, how can destinations enable travel experiences that generate fewer emissions? How can tourism be better designed to address issues of fairness and equity for future generations in an era of climate change?

6.2 **Extending the benefits of tourism**: How might travel experiences motivate travelers to adopt more sustainable lifestyles? How can tourism planners create "staycation" experiences that reduce greenhouse gas emissions, enhance local economies and promote health?

6.3 **Tourists as volunteers**: How can tourists be better engaged as volunteers to restore landscapes, manage and maintain trails, and contribute to small-scale agricultural production and other efforts that restore landscape health? What are the outcomes of those programs for participants, and are those effects durable? Can the flow of benefits and costs of volunteer activity and travel be modeled and optimized for sustainability?

Conclusion

This chapter has explored how tourism systems influence human health, and how human health influences the sustainability of tourism systems. Tourism systems contribute to, and benefit from, the social determinants of health. Health-tourism interactions involve infrastructure and the activities it supports, complex social and ecological relationships and networks, and local economies and employment. In many places where tourism occurs, whether in an informal or highly developed way, the conditions of everyday life are strongly connected to tourism.

Our six-tier framework of tourism–health interactions encourages researchers and planners to contemplate what dimensions of human health they seek to sustain through tourism. The framework highlights many aspects of the social determinants of health, and reinforces why sustainable tourism research focused on human health is needed. It can guide thinking about how tourism influences health for different populations, including within different cultural contexts and landscapes. Successful research programs will likely require multiple disciplines, including landscape architecture, ecology, public health and the social sciences. A consideration of all of the dimensions of this framework will be necessary to build resilience into tourism at all levels, and secure its role in advancing the UN's Sustainable Development Goals. The answers to our proposed research questions will help local communities, governments, NGOs and tourism operators build resiliency by planning and accounting for desired health outcomes for the populations they serve.

References

Bauer, I. (2008). "The health impact of tourism on local and indigenous populations in resource-poor countries." *Travel Medicine and Infectious Disease*, **6**(5): 276–91. https://doi.org/10.1016/j.tmaid.2008.05.005.

Beatley, T. (2017). *Handbook of Biophilic City Planning & Design*. Washington, DC: Island Press.

Boley, B. B. & Green, G. T. (2016). "Ecotourism and natural resource conservation: the 'potential' for a sustainable symbiotic relationship." *Journal of Ecotourism*, **15**(1): 36–50. https://doi.org/10.1080/14724049.2015.1094080.

Bratman, G. N., Hamilton, J. P. & Daily, G. C. (2012). "The impacts of nature experience on human cognitive function and mental health." *Annals of the New York Academy of Sciences*, **1249**(1): 118–36. https://doi.org/10.1111/j.1749-6632.2011.06400.x.

Butler, R. W. (2001). "Seasonality in tourism: issues and implications." In T. Baum & S. Lundtorp (eds.), *Seasonality in Tourism*, pp. 5–21. Oxford: Elsevier.

Cerveny, L. (2008). *Nature and Tourists in the Last Frontier: Local Encounters with Global Tourism in Coastal Alaska*. New York: Cognizant Communication Corporation.

Chason, R. (2018). "Maryland lawmakers move to ensure that all casino funds go to education." *Washington Post*, April 7. Retrieved from https://www.washingtonpost.com/local/md-politics/maryland-lawmakers-move-to-ensure-that-all-casino-funds-go-to-education/2018/04/07/1721fc2e-3a73-11e8-9c0a-85d477d9a226_story.html.

Chen, C. C. & Petrick, J. F. (2013). "Health and wellness benefits of travel experiences: a literature review." *Journal of Travel Research*, **52**(6): 709–19. https://doi.org/10.1177/0047287513496477.

Coghlan, A. & Gooch, M. (2011). "Applying a transformative learning framework to volunteer tourism." *Journal of Sustainable Tourism*, **19**(6): 713–28. https://doi.org/10.1080/09669582.2010.542246.

Commission on Social Determinants of Health (2008). "Closing the gap in a generation: health equity through action on the social determinants of health. Final report of the Commission on Social Determinants of Health," p. 253. Geneva: World Health Organization. Retrieved from http://www.who.int/social_determinants/final_report/csdh_finalreport_2008.pdf.

Connell, J. (2013). "Contemporary medical tourism: conceptualisation, culture and commodification." *Tourism Management*, **34**: 1–13. https://doi.org/10.1016/j.tourman.2012.05.009.

Coon, J. T., Boddy, K., Stein, K., Whear, R., Barton, J. & Depledge, M. (2011). "Does participating in physical activity in outdoor natural environments have a greater effect on physical and mental well-being than physical activity indoors? A systematic review." *Environmental Science and Technology*, **45**: 1761–72. https://doi.org/dx.doi.org/10.1021/es102947t.

Davenport, J. & Davenport, J. L. (2006). "The impact of tourism and personal leisure transport on coastal environments: a review." *Estuarine, Coastal and Shelf Science*, **67**(1): 280–92. https://doi.org/10.1016/j.ecss.2005.11.026.

de Rezende, L. F. M., Rodrigues Lopes, M., Rey-López, J. P., Matsudo, V. K. R. & Luiz, O. do C. (2014). "Sedentary behavior and health outcomes: an overview of systematic reviews." *PLoS ONE*, **9**(8): e105620. https://doi.org/10.1371/journal.pone.0105620.

Deery, M., Jago, L. & Fredline, L. (2012). "Rethinking social impacts of tourism research: a new research agenda." *Tourism Management*, **33**(1): 64–73. https://doi.org/10.1016/j.tourman.2011.01.026.

Flores, A. H., Haileyesus, T. & Greenspan, A. I. (2008). "National estimates of outdoor recreational injuries treated in emergency departments, United States, 2004–2005." *Wilderness and Environmental Medicine*, **19**(2): 91. https://doi.org/10.1580/07-WEME-OR-152.1.

Frith, J. (2014). "History of tuberculosis. Part 2 – the sanatoria and the discoveries of the tubercle bacillus." *Journal of Military and Veterans' Health*, **22**(2): 6.

Fritz, C. & Sonnentag, S. (2006). "Recovery, well-being, and performance-related outcomes: the role of workload and vacation experiences." *Journal of Applied Psychology*, **91**(4): 936–45. https://doi.org/10.1037/0021-9010.91.4.936.

Frumkin, H., Bratman, G. N., Breslow, S. J., Cochran, B., Kahn Jr, P. H., Lawler, J. J., . . . & Wood, S. A. (2017). "Nature contact and human health: a research agenda." *Environmental Health Perspectives*, **125**(7): 075001. https://doi.org/10.1289/EHP1663.

Giles-Corti, B., Broomhall, M. H., Knuiman, M., Collins, C., Douglas, K., Ng, K., . . . & Donovan, R. J. (2005). "Increasing walking: how important is distance to, attractiveness, and size of public open space?" *American Journal of Preventive Medicine*, **28**(2): 169–76. https://doi.org/10.1016/j.amepre.2004.10.018.

Giles-Corti, B. & Donovan, R. J. (2002). "Socioeconomic status differences in recreational physical activity levels and real and perceived access to a supportive physical environment." *Preventive Medicine*, **35**(6): 601–11. https://doi.org/10.1006/pmed.2002.1115.

Haines, A., Kovats, R. S., Campbell-Lendrum, D. & Corvalan, C. (2006). "Climate change and human health: impacts, vulnerability and public health." *Public Health*, **120**(7): 585–96. https://doi.org/10.1016/j.puhe.2006.01.002.

Higham, J., Cohen, S. A., Cavaliere, C. T., Reis, A. & Finkler, W. (2016). "Climate change, tourist air travel and radical emissions reduction." *Journal of Cleaner Production*, **111**: 336–47. https://doi.org/10.1016/j.jclepro.2014.10.100.

Kodish, S. R., Gittelsohn, J., Oddo, V. M. & Jones-Smith, J. C. (2016). "Impacts of casinos on key pathways to health: qualitative findings from American Indian gaming communities in California." *BMC Public Health*, **16**(1). https://doi.org/10.1186/s12889-016-3279-3.

Koetse, M. J. & Rietveld, P. (2009). "The impact of climate change and weather on transport: an overview of empirical findings." *Transportation Research Part D: Transport and Environment*, **14**(3): 205–21. https://doi.org/10.1016/j.trd.2008.12.004.

Kondo, M. C., Han, S., Donovan, G. H. & MacDonald, J. M. (2017). "The association between urban trees and crime: evidence from the spread of the emerald ash borer in Cincinnati." *Landscape and Urban Planning*, **157**: 193–9. https://doi.org/10.1016/j.landurbplan.2016.07.003.

Lacher, R. G. & Oh, C.-O. (2012). "Is tourism a low-income industry? Evidence from three coastal regions." *Journal of Travel Research*, **51**(4): 464–72. https://doi.org/10.1177/0047287511426342.

Lee, C. S., Therriault, D. J. & Linderholm, T. (2012). "On the cognitive benefits of cultural experience:

exploring the relationship between studying abroad and creative thinking." *Applied Cognitive Psychology*, **26**(5): 768–78. https://doi.org/10.1002/acp.2857.

Lee, S. & Jamal, T. (2008). "Environmental justice and environmental equity in tourism." *Journal of Ecotourism*, **7**(1): 44–67. https://doi.org/10.2167/joe191.0.

Li, Q. (2018). *Shinrin-Yoku: The Art and Science of Forest Bathing*. London: Penguin.

Maas, J., van Dillen, S. M. E., Verheij, R. A. & Groenewegen, P. P. (2009). "Social contacts as a possible mechanism behind the relation between green space and health." *Health & Place*, **15**(2): 586–95. https://doi.org/10.1016/j.healthplace.2008.09.006.

Moon, F. (2017). "Five ways to be a savvy medical tourist and enjoy a vacation." *New York Times*, March 8. Retrieved October 4, 2018 from https://www.nytimes.com/2017/03/08/travel/five-ways-to-be-a-medical-dental-tourist-vacation.html.

Morgan, N. & Pritchard, A. (1998). *Tourism Promotion and Power: Creating Images, Creating Identities*. Chichester: Wiley.

Naraindas, H. & Bastos, C. (2011). "Healing holidays? Itinerant patients, therapeutic locales and the quest for health." Special Issue, *Anthropology & Medicine*, **18**(1): 1–6. https://doi.org/10.1080/13648470.2010.525871.

Nguyen, D. B. & Gaines, J. (2017). "Medical Tourism – Chapter 2 – 2018 Yellow Book". Travelers' Health, Centers for Disease Control and Prevention. Retrieved October 3, 2018, from https://wwwnc.cdc.gov/travel/yellowbook/2018/the-pre-travel-consultation/medical-tourism.

Nickerson, N. P., Black, R. J. & McCool, S. F. (2001). "Agritourism: motivations behind farm/ranch business diversification." *Journal of Travel Research*, **40**(1): 19–26. https://doi.org/10.1177/004728750104000104.

NPR, Robert Wood Johnson Foundation & Harvard T.H. Chan School of Public Health (2016). "The workplace and health." Retrieved from https://www.npr.org/documents/2016/jul/HarvardWorkplaceandHealthPollReport.pdf.

Ooi, N., Laing, J. & Mair, J. (2015). "Sociocultural change facing ranchers in the Rocky Mountain West as a result of mountain resort tourism and amenity migration." *Journal of Rural Studies*, **41**: 59–71. https://doi.org/10.1016/j.jrurstud.2015.07.005.

Park, M., Derrien, M., Geczi, E. & Stokowski, P. A. (2018). "Grappling with growth: perceptions of development and preservation in faster- and slower-growing amenity communities." *Society & Natural Resources*, **32**(1): 73–92. https://doi.org/10.1080/08941920.2018.1501527.

Peters, K., Elands, B. & Buijs, A. (2010). "Social interactions in urban parks: stimulating social cohesion? *Urban Forestry & Urban Greening*, **9**(2): 93–100. https://doi.org/10.1016/j.ufug.2009.11.003.

Pizam, A., Reichel, A. & Uriely, N. (2001). "Sensation seeking and tourist behavior." *Journal of Hospitality & Leisure Marketing*, **9**(3–4): 17–33. https://doi.org/10.1300/j150v09n03_03.

Richter, L. K. (2003). "International tourism and its global public health consequences." *Journal of Travel Research*, **41**(4): 340–47. https://doi.org/10.1177/0047287503041004002.

Ruhanen, L. (2013). "Local government: facilitator or inhibitor of sustainable tourism development?" *Journal of Sustainable Tourism*, **21**(1): 80–98. https://doi.org/10.1080/09669582.2012.680463.

Simpson, M. C. (2008). "Community benefit tourism initiatives—a conceptual oxymoron?" *Tourism Management*, **29**(1): 1–18. https://doi.org/10.1016/j.tourman.2007.06.005.

Smith, V. L. (ed.) (2012). *Hosts and Guests: The Anthropology of Tourism*. Philadelphia: University of Pennsylvania Press.

Stokowski, P. A. (1996). *Riches and Regrets: Betting on Gambling in Two Colorado Mountain Towns*. Boulder: University Press of Colorado.

Stonich, S. C. (1998). "Political ecology of tourism." *Annals of Tourism Research*, **25**(1): 25–54. https://doi.org/10.1016/S0160-7383(97)00037-6.

Twohig-Bennett, C. & Jones, A. (2018). "The health benefits of the great outdoors: a systematic review and meta-analysis of greenspace exposure and health outcomes." *Environmental Research*, **166**: 628–37. https://doi.org/10.1016/j.envres.2018.06.030.

United Nations (2017). "Transforming our world: the 2030 Agenda for Sustainable Development."

In *A New Era in Global Health*. New York: Springer. https://doi.org/10.1891/9780826190123.ap02.

United Nations (2018). "68% of the world population projected to live in urban areas by 2050, says UN." May 16. Retrieved October 25, 2018 from https://www.un.org/development/desa/en/news/population/2018-revision-of-world-urbanization-prospects.html.

United Nations World Tourism Organization (eds.) (2010). *International Recommendations for Tourism Statistics 2008*. New York: United Nations.

Urry, J. (1990). "Tourist gaze: travel, leisure and society." *Tourist Gaze: Travel, Leisure and Society*. Retrieved from https://www.cabdirect.org/cabdirect/abstract/19901879617.

US Department of Health and Human Services (2018). "Social determinants of health". Healthy People 2020. Retrieved October 26, 2018 from https://www.healthypeople.gov/2020/topics-objectives/topic/social-determinants-of-health/.

Uysal, M., Perdue, R. & Sirgy, J. (2012). *Handbook of Tourism and Quality-of-Life Research: Enhancing the Lives of Tourists and Residents of Host Communities*. Dordrecht: Springer Science & Business Media.

Vandegrift, D. (2008). "'This isn't paradise—I work here': Global restructuring, the tourism industry, and women workers in Caribbean Costa Rica." *Gender & Society*, **22**(6): 778–98. https://doi.org/10.1177/0891243208324999.

Walker, A. T., LaRocque, R. C. & Sotir, M. J. (2017). "Travel Epidemiology – Chapter 1 – 2018 Yellow Book". Travelers' Health, Centers for Disease Control and Prevention. Retrieved October 3, 2018 from https://wwwnc.cdc.gov/travel/yellowbook/2018/introduction/travel-epidemiology.

White, R. (2011). "Is the staycation trend a real phenomenon?" Kansas City, MO: White Hutchinson Leisure & Learning Group. Retrieved from https://www.whitehutchinson.com/leisure/articles/Staycation.shtml.

Wolf, K. L. & Robbins, A. S. T. (2015). "Metro nature, environmental health, and economic value." *Environmental Health Perspectives*, **123**(5): 390–98. https://doi.org/10.1289/ehp.1408216.

World Health Organization (2015). *Health in All Policies: Training Manual*. Retrieved from http://who.int/social_determinants/publications/health-policies-manual/en/.

World Health Organization (2016). "WHO Constitution." Retrieved October 15, 2018 from http://www.who.int/governance/eb/who_constitution_en.pdf.

11 Public lands, protected areas and tourism: management challenges and information needs

Lee K. Cerveny and Anna B. Miller

Introduction

Protected areas bring people together around common goals of resource conservation and play an important role in catalyzing international dialogue about sustainability (Bushell and Eagles 2006). Tourism is one of the world's fastest growing industries (UNWTO 2018) and there is growing interest in visiting national parks, monuments, forests and reserves for outdoor recreation, wildlife viewing and connecting with nature. Conservation organizations increasingly consider tourism a legitimate vehicle for improving public understanding of natural heritage and elevating the importance of collective conservation action (Fennell and Weaver 2005). Tourism is recognized as a means to achieve biodiversity goals by helping to justify improved planning and management of highly visited protected areas (Becken and Job 2014). Tourism can also invigorate regional economies, alleviate poverty and improve the health and well-being of visitors and nearby residents (Sharpley 2009; see also this volume, Derrien, Cerveny and Wolf, Chapter 10). Sustainable tourism to protected areas can add value to natural and cultural resources, forming a symbiotic relationship, and be congruous with the goals of regional development agencies and indigenous communities.

However, tourism development, increasing visitation, and the rapid diversification of tourist activities can alter natural and cultural resources and create changes in nearby communities (Brockington, Duffy and Igoe 2012). As areas become more heavily visited, deleterious impacts on soils, vegetation, aquatic or wildlife habitat, and water quality can occur and visitors can interfere with wildlife migration and survival (Newsome, Moore and Dowling 2012). Damage to cultural resources and disruption and commoditization of cultural practices is also possible (Greenwood 1989). As visitation expands, it becomes increasingly difficult for managers to keep up with maintenance of built recreation and tourism facilities. Visitor growth leads to increased demands on regional infrastructure, roads and facilities (Mathieson and Wall 1982). The need to accommodate a dizzying array of visitor needs can overwhelm communities, alter the fabric of community life, and affect essential relationships between people and their land (Cerveny 2008; Saarinen 2004). Protected area development also intensifies resource competition and restricts access to certain groups (Agrawal and Redford 2009; Brockington, Duffy and Igoe 2012). Growth in visitation can fuel tensions about land ownership and historic tenure agreements that were altered when the protected area was established (Spence 1999).

This chapter explores management challenges and information needs associated with protected area tourism that can be used to enhance institutional capacity to manage tourism in protected areas. We surveyed 240 resource managers, researchers, tourism proponents, industry officials and environmental organizations with an interest in sustainable recreation and tourism. In this chapter, we share survey results and highlight the research priorities that emerged as five major thematic domains. We conclude by making the link between enhanced research capacity and institutional capacity of resource managers to manage protected areas.

Tourism, public lands and protected areas: a growing synergy

International governing agencies have recognized the need to enhance biodiversity by protecting critical habitat in ecosystems worldwide. According to the International Union for the Conservation of Nature (IUCN), a protected area is "A clearly defined geographical space, recognized, dedicated and managed, through legal or other effective means, to achieve the long-term conservation of nature with associated ecosystem services and cultural values" (Dudley 2008). The Global Protected Areas Program implements the IUCN goals to promote biodiversity conservation, sound governance and public engagement, and ecological resilience against global forces of change. The IUCN classified six major categories of protected areas: strict nature reserves and wilderness areas, national parks, national monuments, habitat/species management areas, protected landscapes/seascapes, and protected areas with sustainable use of natural resources (Dudley 2008). The categories are recognized by the United Nations and by most nations as the global standard for defining protected areas. The Aichi Biodiversity Target 11, which resulted from the Convention on Biological Diversity (CBD), aims to designate 17 percent of the earth's surface as protected parks, refuges, and other designated areas by 2020 (CBD 2011; Chandra and Idrisova 2011). As a result of coordinated efforts, the extent of terrestrial protected areas grew from 8.9 percent of the earth's surface in 1990 to 14.7 percent in 2016 (UNEP and IUCN 2016).

Parks and protected areas have become coveted tourist destinations for the intrepid traveler in search of the wild (Reinius and Fredman 2007). Moreover, the tourism industry has increasingly been viewed as an ally of conservation initiatives, drawing attention, resources and support for protected area management (Bricker, Black and Cottrell 2013). Tourism accounted for roughly 10 percent of the world's gross domestic product in 2016 and is expected to grow 3.9 percent annually for the next ten years (WTTC 2018). Nature-based tourism has been one of the fastest growing tourism sectors (Balmford et al. 2009). Many factors are attributed to protected-area tourism growth, including rising education levels, aging populations, changing gender roles, shifts in leisure time, affordability of travel to distant locales, and growing environmental awareness, among others (Eagles et al. 2002).

Collectively, protected areas receive 8 billion visits annually (Balmford et al. 2015). Visitation levels vary and are higher in more populated areas with natural attractions, such as Europe (3.8 billion) and North America (3.3 billion) and lowest in Africa and Latin America. Visitation to protected areas in the developed world is level or steady, while visitation to newly established protected areas in developing regions may be growing faster than average (Karanth and DeFries 2011). The associated $600 billion in visitor spending (Balmford et al. 2015) represents important economic opportunities for host nations and potential sources of revenue to support the system of protected areas to meet conservation goals. Parks and protected areas are increasingly viewed as a source of revenue to host countries and regions (Sharpley 2009). These revenues are often used to support and improve transportation systems, infrastructure and community services. Tourism generates direct employment opportunities in industries like guiding, hospitality, retail, and sale of handicrafts; and indirectly in the production of goods and services that support the tourist trade. Generating local economic opportunities is important for ensuring long-term community support for protected areas, especially when their establishment has reduced local access (Job and Paesler 2013).

Tourism, when managed sustainably, can improve the quality of life for inhabitants and protected area visitors (McCool 2006). Parks and protected areas provide ecosystem services, benefits to people in the form of clean air, clean water, carbon storage, food, provisions, and health benefits, among others (Frumkin et al. 2017; Watson et al. 2014; see also this volume, Derrien, Cerveny and Wolf, Chapter 10). However, tourism to protected areas can also lead to unwelcome changes in communities, due to the uneven distribution of economic benefits, commoditization of customs and traditions, and an influx of new workers, non-local businesses, and in-migrants (Andereck et al. 2005). Conflicts can arise when benefits do not transpire for those investing in tourism or seeking employment (Dhakal, Nelson and Smith 2011). Tourist visitation can put pressure on local infrastructure or create competition for limited resources (West, Igoe and Brockington 2006). While protected area tourism can expand the constituency for biodiversity and conservation efforts, visitor growth can sometimes bring lasting changes to resources and people, not all of them positive.

Escalating visitation to protected areas also can result in challenges for protected area managers (Eagles 2014). Visitors demand improved facilities and services and may arrive seeking new outdoor activities or with technology that changes their visitation patterns, leaving some land managers unprepared (Shultis 2001). In some areas, high visitation levels may reduce the quality of the visitor experience for those seeking solitude or whose use is in conflict with others (Floyd 2001). Land managers seek help in providing quality visitor experiences while protecting natural and cultural resources. They need frameworks and concepts that explore visitor motivations, expectations and responses to settings provided. Meanwhile, in regions with low or modest visitation, land managers seek ways to better understand how to generate visitor interest, align protected area goals with the interests of tourism promoters and local officials, and market the protected area.

Sustainability science encourages tourism development that "takes full account of its current and future economic, social and environmental impacts, addressing the needs of visitors, the industry, the environment and host communities" (UNWTO 2005). This suggests planning protected area tourism to maintain essential ecological processes and conserve natural heritage and biodiversity, acknowledge the authenticity of host communities and cultural heritage, and encourage tourism activity that results in lasting economic operations, with evenly distributed economic benefits, job opportunities and poverty reduction strategies (UNWTO 2005). Today, public lands and protected areas are being impacted by dramatic changes, including: (a) climate and weather patterns leading to flooding, drought, wildfire, higher temperatures and other risks; (b) human population growth, leading to resource competition, development pressure and increased energy needs; (c) global economic shifts that impact transnational trade regulations, corporate investment, infrastructure and utilities development, and labor markets; (d) war, conflict and changes in border policies that affect who can travel and where they can go; (e) human disease, famine or safety, which affect travel motivations and experiences; and (f) shifts in environmental values which affect how people view humans' role in stewarding our planet. Sustainable tourism practices are more important than ever to encourage resilience in ecological and socio-cultural institutions and processes.

Managing tourism in protected areas, particularly in lieu of global changes, requires new sources of knowledge and information that look critically at what protected area tourism should sustain (Berno and Bricker 2001). With attention to national parks, Eagles (2014) identified areas for future research leading to improved visitor management and monitoring, enhanced visitor satisfaction, financial benefits and good governance. While these are important areas, more work is needed. Research on the tourism sector will help to better understand the changing demographics, consumer trends and technologies that drive tourist activity to help managers better predict visitor behaviors and resource interactions. New knowledge about visitor experiences and settings will allow better design of facilities and programs. Improved understanding of the relationships between visitor use and impacts will enhance managerial capacity to match tourism development with existing capability. Understanding the implications of park visitation on local populations allows more equitable and socially sustainable solutions. Research plays a significant role in building capacity for effective protected area management.

Study approach

To better understand research and information needs in sustainable recreation and tourism for protected areas, we conducted a needs assessment survey in 2017–18 of resource professionals and researchers worldwide. Questions focused on perceived management challenges, information and research needs to address challenges, and ideas about how tourism can be managed sustainably. The survey included a mixture of fixed answer responses and open-ended questions to elicit ideas about the pressing challenges facing today's protected area managers.

An iterative snowball sampling approach was used to ensure that the survey was accessible to a wide range of prospective respondents. The study team initially identified 99 people through professional networks and encouraged people to forward the survey to others in the field. To expand the perspective, the survey was distributed to several professional networks, including the IUCN's Tourism and Protected Areas Working Group, a cultural heritage management group, a recreation ecology professional network, and to participants in a workshop on recreation and wildlife. Each of these networks included hundreds of participants. The study team conducted an initial round of qualitative data analysis, which established a set of thematic codes. The data then were reanalyzed using the thematic codes, and responses were tallied and prioritized using the codes to determine the highest priority management challenges and information needs.

We received 240 responses. Because there is no way to determine how many of the surveys were forwarded or to whom, we did not compute a response rate. More than half of the respondents were affiliated with governmental land management agencies (53%). Other responses came from universities (25%), non-governmental organizations (15%), consultancies (5%), and the private sector (3%). The majority of respondents work on resource management primarily in the United States. (86%), while at least 18 percent work outside of the United States. Respondents occupied a variety of professional roles in the recreation and tourism sector and many occupied more than one role. More than half (55%) identified as recreation or tourism planning professionals, 44 percent were in research or academia, 39 percent were policy or decision makers, 38 percent were land managers, 28 percent were involved in environmental assessment, 27 percent engaged in tourism capacity development, 17 percent in science communication, and 11 percent in technology development.

Management challenges, information needs and research directions

Priority management challenges relating to public lands and protected areas are presented below based on our survey responses. We identified five broad domains that collectively contain 18 themes that capture the priority information needs and frame the need for future research on public lands and protected areas. We acknowledge regional variations in the relative importance of each theme. Since a predominance of our respondents were based in North America, domains likely reflect challenges faced in those settings. The domains are: (1) social-ecological interactions, (2) visitation trends, (3) visitor experiences and benefits, (4) community and culture, and (5) governance, planning and sustainability monitoring. Below we describe each domain and the critical themes that characterize each. When appropriate to illustrate the need, we include direct quotations from respondents. Taken together, these five domains reflect an encompassing view of sustainability with regard to protected area tourism, although it is certainly not a complete picture.

Domain 1: Social-ecological interactions

A social-ecological systems approach recognizes the human dependence on protected ecosystems and how changes in those systems affect people. Conservation problems often occur when land management agencies and policies fail to recognize the interdependence of social, ecological and economic systems in protected areas (Cumming and Allen 2017).

1.1 *Ecological effects*

Heavy tourism in protected areas, when not properly managed, can lead to biodiversity loss and alter ecosystem functioning (Newsome, Moore and Dowling 2012). As visitors walk along trails, their footprints can damage plants and compact soils, leading to increased erosion along trail pathways. Compacted soils, erosion and disturbed native plant communities, combined with invasive plants hitchhiking on tourists' clothing, can alter the vegetation community and degrade water quality. Recreational infrastructure such as trails, campsites or other cleared areas, create edge habitat, contributing further to ecosystem shifts and biodiversity loss. Such changes can contradict protected area missions, particularly for protected areas designated to conserve a threatened ecosystem (Marion et al. 2017).

One manager wondered, *"How [do we] manage recreation demand while adequately protecting forest resources from damage/overuse with limited resources?"* Early research on the effect of people walking on vegetation, simulating off-trail hiking, indicated that low levels of human use had disproportionately high impact on the vegetation, and that this impact leveled off at a certain amount of use (Cole 1981; Hammit, Cole and Monz 2015). This use–impact relationship has been generalized to ecosystems and biophysical factors to which it does not apply, influencing management strategies in protected areas in many parts of the world (Monz, Pickering and Hadwen 2013). Further research is needed to develop our knowledge of the ecological effects of protected area tourism, taking into account biophysical factors such as topography, vegetation type, and substrate. Understanding and planning for user demand using social science works towards solving issues related to unintended park use, such as user-made trails that can harm natural resources, pose safety issues for tourists, and present maintenance issues to management. Research needs identified by study participants include:

1. Understanding use–impact relationships on a wide range of biophysical factors and ecosystems, and applying this knowledge to visitor management.
2. Using a social-ecological systems framework to manage protected area tourism.

3. Understanding tourist demand to alleviate problems related to visitation growth.

Integrating research on tourist demand with an improved understanding of use–impact relationships can contribute towards more sustainable protected area management. Knowledge produced through this research will enable managers to identify areas where tourism is more or less likely to have deleterious effects on protected ecosystems, and make informed decisions on where to focus tourism activity and how to allow safe tourism in sensitive areas.

1.2 Human–wildlife interactions

Protected area tourism and associated infrastructure can lead to habitat loss, displace animals from preferred habitat, affect breeding success, disrupt movement between critical habitat areas, and result in vehicle collisions causing injury or death, among other impacts (Hammit, Cole and Monz 2015; Knight and Gutzwiller 1995). The effects of recreation vary, with countless combinations of environmental, wildlife, and human variables. As one manager explained, "*We struggle with the lack of science around the different recreational impacts on wildlife and trying to keep it in context of other impacts that already exist.*" Despite extensive research on this subject, few general patterns have emerged, especially at the population level, and many studies have presented contradictory results (Bateman and Flemming 2017; Larson et al. 2016; Marion et al. 2017; Tablado and Jenni 2017). New recreational activities and demands outside of the developed recreation infrastructure add to the complexity of human–wildlife interactions. While many studies indicate negative effects of recreation to wildlife at the individual level, in the short-term, and at relatively small spatial scales, fewer studies have drawn conclusions regarding how these translate to long-term population-level effects across broad areas (Tablado and Jenni 2017).

Human–wildlife conflict can pose safety issues for tourists, threaten residents' livelihoods and negatively impact wildlife populations. It is critical to consider characteristics of both wildlife and humans within the system (Lischka et al. 2018). Understanding the assumptions used to form mental models of a human–wildlife interaction system can aid in problem-solving (Mosimane et al. 2013). Research in the following areas will contribute towards our understanding of human–wildlife interactions:

1. Apply and develop theoretical frameworks to extrapolate short-term individual-level effects to long-term population-level effects, to elucidate ways in which protected area tourism affects wildlife activity in lasting ways, and how negative impacts can be avoided.
2. Landscape-scale research across jurisdictions and ecosystems is needed to provide a more complete understanding of human–wildlife interactions, since wildlife habitat crosses man-made boundaries and human developments cross multiple habitats.

3. Integrated research accounting for human, wildlife and environmental variables is needed to inform management.

Such studies would contribute greatly to our understanding of recreation–wildlife interactions. Research in this area will contribute towards minimizing negative human–wildlife interactions and build more symbiotic relationships between humans and wildlife in protected areas.

1.3 Environmental shifts and disturbance factors

As we experience the effects of climate change and altered fire regimes, researchers are beginning to investigate how these large environmental events and shifts are affecting tourism in protected areas (Scott et al. 2008). One manager highlighted the "*huge need related to climate change and the potential impacts to recreation.*" In areas where temperatures are rising, the snow-free tourism season is expected to lengthen, while the snow-based tourism season will shorten (Hand et al. 2018). Longer shoulder seasons will likely exacerbate managerial issues in protected areas which already struggle to meet existing tourism demands. Wildfire can also affect tourism to protected areas (Sanchez, Baerenklau and Gonzalez-Caban 2016). Other environmental changes, such as the spread of invasive species, may cause shifts in recreation and tourism. Restoration efforts also may alter recreation patterns, as areas are treated using controlled burns, salvage logging or other activities. Research on ecological disturbance effects on tourism is in relatively early stages, and has started with a focus on the associated economic impacts. Further research is needed to better understand the relationship between environmental shifts, tourism and human communities, including:

1. Understanding tourist substitution patterns and choices under altered climatic conditions.
2. Understanding the effects of wildfire, invasive species and other disturbances on tourism activities.

A better understanding of the relationship between environmental shifts and protected area tourism will prepare managers for altered visitation trends, an expression of shifting demand and availability of protected areas for tourism.

Domain 2: Visitation trends

Understanding visitation to protected areas is important for the continued support of protected area conservation and management of natural, cultural and managerial resources. Much of the existing research on protected area tourism has focused on visitor numbers and management tools (Buckley 2012; Eagles et al. 2002). In some protected areas, particularly in the developed regions, the focus lies on understanding and managing high visitor volumes, whereas in other places, including developing regions and newly designated parks, emphasis is on generating visitation, leading to increased support for conservation. Understanding visitation through measuring and monitoring visitor use allows managers to minimize potential impacts to social and natural resources (McCool and Spenceley 2014).

2.1 *Managing visitor volumes*

In some protected areas, a high volume of visitors can lead to perceptions of crowding or diminished quality of the recreation experience (Buckley 2009). Planning frameworks, such as the Recreation Opportunity Spectrum (Clark and Stankey 1979), carrying capacity (Graefe, Vaske and Kuss 1984), and Limits of Acceptable Change (Stankey et al. 1985) have been incorporated into policy for some protected areas, but implementation is criticized and these tools are applied inconsistently (Cerveny et al. 2011; McCool, Clark and Stankey 2007). Additionally, new ways of sharing information have altered visitation patterns. The use of social media has generated new demand for sites that are photo-tagged and shared, sometimes resulting in surges of visitation (Wood et al. 2013). In the United States, the Interagency Visitor Use Management Council has brought together six federal agencies to develop a framework for a flexible process for managing visitor use in protected areas (IVUMC 2016). Ten principles for visitor management were recently described within an IUCN Best Practices Guidelines document, with the objective of improving effectiveness and increasing public support for visitor management (Leung et al. 2018). Survey results suggest that research on visitor management should focus on:

1. Assessing the efficacy of existing management principles to inform visitor management.
2. Developing and evaluating new approaches for managing high-volume visitation.
3. Developing new methods to understand visitor demand in real time, such as using social media data to predict patterns of recreation use.

Research in this area will contribute towards visitor management adapted to changing visitation trends in areas with a high volume of visitation. By enhancing our understanding of protected area visitors and their use patterns, managerial capacity to address impacts of visitation on natural and cultural resources may be expanded.

2.2 *Generating visitation to protected areas*

Some protected areas face issues due to a lack of visitation, resulting from structural or cultural barriers. One respondent stated, "*In developing countries, the protected areas agencies don't know how to attract more people to the parks and the majority remains with hardly any visitation at all.*" Structural barriers stem from a lack of infrastructure, and can occur locally, regionally or internationally. At the park level, a lack of facilities or insufficient information about existing resources can make tourism difficult. Absence of roads or transportation to access the protected area can also result in reduced visitation (Leung et al. 2018; Spenceley 2008). International tourism to a protected area can be inhibited when travel information is unavailable, local transportation systems are under-developed, or political factors or local practices (e.g., poaching) raise safety concerns. Cultural barriers can also preclude tourism to protected areas, particularly in areas where being outdoors is not considered a leisure activity. Depending on the financing structure, such areas may also face severe maintenance or staffing backlogs. Survey responses indicated that sites with low visitation need research to:

1. Identify structural and cultural barriers to protected areas and methods to overcome the identified barriers.
2. Present site-specific case studies to identify and overcome barriers.
3. Explore the potential of social marketing strategies to identify potential visitors to the protected area and incentives to encourage their arrival.

Research in this area will aid protected areas which face low visitation with strategies to gain public support for the sustainability of these areas through increased visitation.

2.3 *Monitoring visitors*

To understand protected area visitation trends, monitoring programs are carried out by many land management agencies. Existing programs can be expensive to implement and often take a lower priority than other programs, resulting in incomplete data. One respondent indicated, "*We don't know much about how many people frequent which forests, what they do there, etc. . . . However, this information is indispensable for effective visitor management.*" When visitor use data are collected, the methods vary by spatial and temporal scale, level of detail, and often lack site-specific information on economic and sociocultural impacts. These inconsistencies limit integration of visitation data between sites managed by different agencies. This becomes problematic in understanding how large-scale environmental shifts affect protected area tourism across multiple scales. Visitor monitoring research has recently turned towards using data acquired through social media as a cost-effective way to gather standardized data at fine-grain spatial and temporal scales (Fischer et al. 2018; Walden-Schreiner, Leung and Tateosian 2018). Efforts are also being made to share data on protected area visitation at a global scale (Schägner et al. 2018). Survey results indicated that research to improve visitor monitoring should seek to:

1. Develop a standardized, adaptable, and cost-effective method for collecting visitor use data, including the use of social media data, when appropriate.
2. Focus new tools on gathering visitor data at spatial and temporal scales relevant to management.

Improved visitor monitoring methods will help researchers understand how environmental events affect recreation at the landscape scale, and assist managers with understanding and managing changing visitation trends within the context of change. More accurate visitor counts can justify the continued support of protected areas to decision makers.

Domain 3: Visitor experiences and benefits

The visitor experience has long been a component of protected area planning and management. The notion that people visit protected areas seeking a certain experience is built into many recreation planning frameworks, such as the Recreation Opportunity Spectrum (Clark and Stankey 1979), Limits of Acceptable Change (Stankey et al. 1985) and others. Driver and Brown (1978) described a four-level hierarchy of recreation demand based on increasing levels of complexity, starting with recreation activities as the least complex, followed by settings, experiences and, finally, benefits. Protected area visitors bring preferences to their trip and managers provide opportunities from which visitors choose, allowing visitors to produce their desired experiences (Moore and Driver 2005). A growing body of research focuses on recognizing benefits of protected areas to visitors, including health, well-being, spiritual, sustenance, social and economic (Bryce et al. 2016; Frumkin et al. 2017; Twohig-Bennett and Jones 2018). Some land management agencies, including the U.S. Bureau of Land Management, use a benefits-based approach to management, monitoring indicators of personal, community and social, environmental, and economic benefits.

3.1 Understanding visitors and their needs

Who is coming to our parks and protected areas? What societal forces frame their visit? What motivations, values and perceptions do they bring? Visitors to public lands and protected areas are constantly changing along with shifts in demographics, residential status (urban/rural), socio-economic status and work/leisure patterns (McDonald et al. 2009; Weiermair and Mathies 2004). In the past, travel to public lands and protected areas was only possible for the wealthy. As travel became more affordable and public lands become more accessible, new visitors ventured out, bringing different interests and attitudes towards the environment (Berno and Bricker 2001). In many nations, the population is aging, suggesting different kinds of visitor experiences, namely a shift from active to passive recreation and from 'backcountry' to 'frontcountry' settings. In industrialized nations, ideas about work and leisure have been blurring and transitioning, with shorter vacations and vacations where work and other obligations are interspersed with leisure. In North America, visitors to protected areas are spending less time, preferring day-use to overnight stays, and are concentrating use close to roads and developed areas (White 2016). In many places telecommuting makes it possible to live near public lands, which allows for frequent interaction with nature. All of these elements are changing the nature of the visitor experience. One respondent indicated that "*better understanding how people connect to nature and why is a topic that needs attention.*" Survey respondents identified the need for research in the following areas:

1. Understanding who is visiting protected areas and how visitor demographics and motivations might be changing.
2. Understanding visitors' ideas, attitudes, constraints, and intended interactions with protected areas.

Knowledge gained from research in this area will help protected area managers anticipate changes in visitation and be prepared to accommodate a more diverse group of visitors.

3.2 *Managing visitor experiences*

Visitor experiences are generated by the visitor who interacts with the natural and social environment within the realm of the opportunities and settings available in a protected area (Moore and Driver 2005). In this sense, visitor experiences are managed by those providing the protected area settings, and mediated by the visitor through the ways in which he or she interacts with the site. In addition to research on the effects of settings, opportunities and experiences, there is an explicit need to better understand the nature of the experience itself. One respondent explained, "*To date we have focused on activities, now we need to move a level deeper into why people are engaging in those activities and what outcomes they are receiving from those experiences.*" Many recognized that technology, including both information technology and recreation technology, mediates the recreation experience (Pohl 2006; Pope and Martin 2011). Prospective visitors planning their trip from home are shaping their expectations based on what they have learned about the destination online and what their friends have said about the site or the trip on social media (Wilkins, Smith and Keane 2018). Once they arrive, visitors may interact with the protected area using personal devices such as phones that interpret the experience for them in real time. Survey results suggested that visitor experience management could be improved by research in the following two areas:

1. Developing new methods for understanding the experiences desired by protected area visitors.
2. New frameworks for determining how desired tourism experiences are met in the protected area setting.
3. Understanding the role of information technology in visitors' protected area experiences.

A greater understanding of the visitor experience will help park and protected area managers to shape expectations, provide appropriate visitor services and programs, predict visitor activities and actions, and plan for high quality experiences in a variety of settings.

3.3 *Providing diverse experiences to diverse visitors*

An increasingly culturally, economically and socially diverse pool of visitors places new expectations on managers. Moreover, as the health benefits of nature-based recreation are being realized, there is a moral and ethical responsibility for nations to expand these benefits to all of their inhabitants. Yet, protected areas are not always accessible to all socio-economic groups, and some areas have excluded the local populations who once depended on the now-protected lands for sustenance and survival. Barriers to access vary widely, and examples include lack of transportation, lack of opportunities available which are in demand by different socio-cultural groups, high price of entry, parking fees or other expenses, and distrust of the managing organization (Buckley 2003; Dawson, Martin and Danielsen 2018). Equity of access is important for both ethical and practical reasons, to avoid conservation causing negative impacts to local populations, and to increase the effectiveness of conservation (Hutton, Adams and Murombedzi 2005). One respondent explained, "*If people from different minorities do not see themselves represented in outdoor recreation, and if marketing/outreach efforts are not tailored specifically to those populations, then these people will not see outdoor recreation as inclusive or enriching.*" Survey results suggested the need for research in the following areas:

1. Improving the understanding of barriers that people face in accessing and participating in outdoor recreation and tourism in protected areas.
2. Developing new communication strategies for reaching diverse visitors to understand unique visitor needs, share information, and to provide a high quality experience.
3. Understanding how to better provide diverse experiences within protected area settings for a variety of user types.

This subject echoes the recurring topic of understanding elements of demand for protected area tourism. New knowledge in this area will help protected area managers expand access to opportunities in protected areas to a more diverse group of visitors.

3.4 *Maximizing benefits to visitors*

Benefits of protected areas to visitors come in many forms, such as health and well-being, education, connection with nature, stewardship, spiritual and economic. While benefits can be difficult to measure, some have been incorporated into protected area management programs. Deriving benefits from protected area tourism is linked with positive outdoor experiences (McCool 2006). As one respondent explained, "*The way these benefits are expressed and valued may differ from place to place, but at the base of it is a human expectation that our global network of protected areas will literally save us from ourselves.*"

Research efforts to better connect recreation and tourism to the flows of benefits that people derive from public lands and protected areas have taken place in the context of ecosystem services (Millennium Ecosystem Assessment 2005). This framework conceptualizes a broad array of benefits that include cultural ecosystem services, incorporating non-market benefits such as spirituality, scenery, learning, heritage and stewardship. While cultural ecosystem services can be difficult to operationalize and measure, promising research is finding ways to quantify and demonstrate these benefits for protected areas (Watson et al. 2014). Two areas of inquiry were recommended:

1. Interdisciplinary research to identify and understand benefits derived from protected areas.
2. New methods for assessing benefits and incorporating them into management programs.

Finding ways to understand and communicate benefits of tourism to public lands and protected areas will help to build capacity for long-term management.

Domain 4. Community and culture

Tourist visitation to protected areas affects the lives of indigenous people and residents of gateway and 'host' communities (Espiner and Becken 2014; Strickland-Munro, Allison and Moore 2010). Sustainable tourism approaches seek to maximize tourism benefits to local residents while minimizing or mitigating unwanted effects. Community engagement in tourism development is socially responsible and leads to higher quality visitor experiences (Jurowski and Gursoy 2004). Visitation occurs on lands important to local and indigenous communities for resource uses, economic opportunities, lifeways and cultural practices. Resource managers seek innovative and inclusive approaches to engage residents in planning and management and to acknowledge indigenous treaty rights through co-management. Protection of cultural resources is a critical element to sustainable tourism management.

4.1 *Tourism benefits and challenges for communities*

Public land managers seek strategies that promote community well-being and resilience to external and emergent processes that ignite change. Tourism impacts on communities have been well-documented globally (Sharpley 2014). Tourism can lead to changes in community or rural character, quality of life, in-migration of tourism workers, concerns about commoditization and consumption of identity, resource conflict, authenticity, protection of indigenous traditions, and cultural resources (Smith and Brent 2001). These issues can be exacerbated when agencies and stakeholders in the tourism system are not engaged in coordinated planning and management. Protected areas and the tourism associated with them may provide tangible benefits to host communities that have not been fully realized. One respondent suggested, "*Recreation is an asset. Rural communities need to look at it that way and land managers need to support rural place-making in ways that positively contribute to the agency's goals to be a good neighbor while allowing for sustained use.*" Protected area tourism provides a unique case, given that some visitors may not set foot into local communities. Tourism growth to public lands can ignite other changes, such as amenity in-migration, development, and conversion of resource-based industries to tourism (Rebollo and Baidal 2003). Community-based tourism plans are considered an effective means of increasing local leverage. By promoting opportunities for community and stakeholder engagement in planning, local capacity for tourism management is also expanded beyond the public agency. Survey results suggest four themes:

1. Help communities to assess the potential benefits of protected area tourism and develop approaches to expanding community roles in the tourist enterprise.
2. Develop tools to assess tourism benefits and costs, and consider their equitable distribution among socio-economic groups and stakeholders.
3. Understand direct and indirect processes of development associated with protected area tourism leading to community change.
4. Identify strategies for expanding community engagement in tourism planning.

Research to understand the depth and diversity of tourism effects and how these implicate a variety of stakeholders may result in decision tools that ultimately help managers to minimize effects and communicate challenges.

4.2 *Understanding economic benefits of tourism*

Carefully planned tourism can be an important economic development opportunity for rural communities that serve as gateways to protected areas (Eagles 2002). Sustainable tourism promises to bring economic benefits to host communities, although the jury is out on whether these benefits are fully realized (Stronza 2007; Wunder 2000). Tourism also can result in employment in auxiliary sectors such as transportation, agriculture and manufacturing, which produce goods to support the industry. In many regions, tourism is considered an appropriate tool for poverty alleviation (Adams and Hutton 2007; Ferraro and Hanauer 2011). Managers seek information about how communities can maximize socio-economic benefits of protected area tourism through investment, how to measure and communicate economic benefits, how local industries and individuals can increase investment in tourism, and the extent to which tourism is helping the poor. Approaches that help managers identify tourism beneficiaries as well as those not participating in the tourism economy, and understand tourism marketing and distribution systems and their effect on the flow and volume of visitors to the park, would be beneficial. As one respondent noted, "*We need to understand the economic importance of tourism. How much money directly and indirectly flows through a local community because of the tourist activity?*" This knowledge would be enhanced by strategies to ensure that the tourism benefits are distributed fairly and do not remain in the hands of a few established firms. Research needs identified by study participants include:

1. New frameworks to maximize economic opportunities and reduce poverty for adjacent communities.
2. Understanding global–local interactions among tourism providers and impacts on park visitation, local economic opportunities and tourism markets.
3. Measuring tourism economic impacts and mapping tourism beneficiaries and non-beneficiaries at the local and regional level.

General knowledge of the economic structure of the tourism industry would be valuable to protected area managers to understand the effect of global economic activities on local transactions.

4.3 *Engaging indigenous leaders in tourism planning*

Indigenous peoples are increasingly assuming leadership of tourism enterprises as a way to control the terms of tourism, proclaim authority of indigenous history and cultural practice, provide benefits to their communities, and to expand awareness of their concerns (Butler 2007). There is growing recognition that tourism can be an appropriate vehicle for growing indigeneity (Carr, Ruhanen and Whitford 2016). Parks and protected areas often are located in areas valued by indigenous peoples. For tourism to be sustainable, protected area managers recognize the role of indigenous groups with legal rights and enduring land tenure relations and use patterns. Protected areas are forging new relations and finding ways to engage indigenous leaders and critical stakeholders in park management (Fletcher, Pforr and Brueckner 2016). In our survey, we observed the need for diverse engagement of indigenous leaders in tourism planning processes, with emphasis on expanding that diversity to traditionally under-represented groups, whose voices are not always heard. Several respondents mentioned the need to work with tribes and develop approaches for, *"fostering indigenous reconciliation and indigenous-led conservation in an ethical way."* Survey respondents identified research needs in the following areas:

1. Empowerment models of tourism planning in protected areas that engage indigenous leaders and promote expanded indigeneity.
2. Mechanisms for mediating effects of tourism on local infrastructure and institutions through public–private partnerships and collaboration.
3. Mechanisms for governing local access through adaptive co-management to promote shared conservation goals.

Appropriate community-based planning efforts that engage stakeholders, integrate local and traditional ecological knowledge, recognize historic use patterns, and give voice to historically under-represented groups would contribute to sustainable management of protected area tourism.

4.4 *Protecting and enhancing cultural resources, properties and traditions*

Increasing human visitation to protected areas can have deleterious effects on cultural heritage resources when not managed effectively (Timothy and Nyaupane 2009). Partnerships between indigenous groups, tourism promoters, resource managers and cultural heritage professionals can help to identify strategies for minimizing impacts to cultural resources and increasing awareness of the importance of authentic cultural experiences (McKercher and Du Cros 2002). There was a demonstrated need for understanding the impacts of protected area visitors on cultural resources. As one respondent explained,

> *I find that the agency does not spend near enough time, effort, and funding in gathering data to understand the impacts of outdoor recreation on cultural resources and tribal resources. Both resources have been severely impacted by recreation, yet there is very little*

hard, baseline data established, or coordination and consultation with the Tribes to understand what impacts may be occurring or where they are occurring.

Resource managers should consider authenticity in the interpretation of natural and cultural sites and ask themselves, 'Whose story is being told here?' In many cases, cultural connections are overlooked in areas valued for their natural heritage. Research needs include:

1. Approaches that promote appreciative cultural interpretation that encourage resource protection.
2. Strategies to prevent damage to cultural resources while encouraging tourism visitation to cultural sites.
3. Interpretive strategies that recognize the multiplicity of voices and interpretations of natural and cultural heritage.

With increasing visitor numbers, cultural areas and materials face ongoing threats. The need for stronger relationships with indigenous groups were noted.

Domain 5: Governance, planning and sustainability monitoring

Worldwide, protected areas have benefitted from a global system of governance that has provided consistent standards for measuring outcomes (Bricker, Black and Cottrell 2013). There is growing recognition of the importance of governing agencies to coordinate across jurisdictions to meet mutual conservation objectives; to engage local, indigenous, regional, and global actors and address stakeholder concerns; to contribute to regional economic vitality and reduce poverty; and to manage public lands and protected areas with a focus on increasing community, institutional and ecosystem resilience. Many protected areas face increasing tourism demand while capacity for management is declining. Protected area management requires updated methods and indicators to holistically assess whether actions are meeting intended goals and to allow potential for adaptation to constantly changing systems (Graham, Amos and Plumptre 2003). Creative funding sources and financing strategies will help ensure that tourism enterprises and visitors contribute to the long-term viability of protected areas, increasing the likelihood that conservation goals can be met.

5.1 Sustainable tourism monitoring

There is widespread recognition about the need to develop a standard index that can be adapted and applied across protected areas that would consider ecological, economic, and socio-cultural components while accounting for local conditions (Bricker and Schultz 2011). Work to develop consistent guidelines and operating procedures across protected areas globally is in its early stages (Bushell and Bricker 2017). One respondent stated,

> *We need an overarching sustainability framework . . . with a focus on the institutional (governance) mechanisms to pull the classic aspects of sustainability (economic, environmental and social) towards some positive aspect . . . while considering the rapid/changing needs and interests among the publics.*

Many indices of sustainable tourism have been developed, discussed and implemented in a variety of settings. Efforts have been made to adapt these indicators to local conditions with varying degrees of success (Lee and Hsieh 2016; Tanguay, Rajaonson and Therrien 2013). Studies have found that socio-economic and environmental indicators can help objectively measure the degree of sustainability (Choi and Sirikaya 2006; Singh et al. 2009). However, there is less agreement about how individual indicators contribute to sustainable tourism goals or the relative weight of indicators. Efforts to compile a global database to report on several indicators of protected area visitation, including tourism use, tourism value, and tourism-related economic impacts of protected areas, are underway and may offer guidance (Schägner et al. 2018). Public land managers are eager to employ an approach that has been tested and that is implementable within the fiscal and operating constraints of public agencies and in concert with other assessment tools. According to our survey responses, research needs in this area include:

1. Assessment approaches that are adaptable to national priorities and local conditions, coupling natural and socio-cultural systems.
2. Understanding the impact of indicators, their relative weight, and how they promote sustainability.
3. Identifying tools that are affordable, implementable, and that measure sustainability on multiple realms.

Sustainability indicators that can be adapted to multiple settings and scales will help resource managers across jurisdictions develop shared goals, establish a common language and gain practice and professionalism toward improved visitor management. Criteria that embrace social, cultural, ecological, economic and governance factors will allow a holistic understanding.

5.2 *Effective governance*

Tourism to protected areas has historically been managed by single institutions or agencies within a nation reflecting distinct agency goals (e.g., preservation, multiple use). More recently, protected areas are understood as a global system organized to achieve biodiversity conservation. Global strategies for conserving nature through protected areas have been effective in developing consistent management practices (Bushell and Bricker 2017). Still, the actual management of the protected areas often falls within a particular agency's purview and is subject to the whims of national leadership. One respondent noted, "*The biggest challenge is capacity – the capacity for the protected areas to support human recreation; the capacity for agencies to deliver recreation opportunities.*" Eagles (2009) reviewed multiple studies to identify essential governance characteristics for protected area tourism and identified critical factors, including public participation, consensus orientation, strategic vision, responsiveness to stakeholders, effectiveness, efficiency, accountability, transparency, equity and law. Protected area managers seek information about innovative systems for managing visitation across borders, jurisdictions and sectors. Collaborative governance can "engage people constructively across the boundaries of public agencies, levels of government, and/or the public, private and civic spheres in order to carry out a public purpose that could not otherwise be accomplished" (Emerson, Nabatchi and Balogh 2012: 1). Plummer and Fennell (2009) describe the value of adaptive co-management for protected area tourism. One respondent stated the need for *"collaborative approaches that can improve management of protected areas through open-source sharing of best practices, research, and other relevant materials and also by providing a platform for scaling up our collective influence and impact."* Respondents identified three research themes:

1. Better understanding of the conditions of land ownership, degree of sector engagement (public, private, civic), management agency authority, and financial capacity.
2. New approaches to encourage inter-agency coordination around tourism at multiple spatial scales.
3. Exploring models for public–private partnership, community and tribal investment in parks and protected area management.

Public lands governance can no longer be achieved by singular agencies alone, particularly in light of declining capacity, including personnel, budgets, information and infrastructure. New approaches are needed to integrate interests of stakeholders and institutions at multiple levels.

5.3 *Innovative funding models*

Currently, most protected areas are funded by individual governments, and in many instances budget declines have impacted agencies' ability to adequately steward the land and uphold conservation goals. Tourism has recently been viewed as a mechanism to finance the growing system of protected areas (Karanth and DeFries 2011). Eagles (2014) draws attention to the need for new and innovative park financing strategies, including visitor fees, licenses and permits, rental fees, concessions contracts, and others, noting the potential for public–private partnerships and community-based strategies. Whitelaw, King and Tolkach (2014) review alternative financing strategies for different types of protected areas, based on varying visitation levels and degrees of biodiversity. One agency official suggested a need for *"sustainable and long-term dedicated funding mechanisms for public land recreation . . . and public/private partnership models that encourage private investment with some level of reasonable return."* Others suggested that protected areas draw from an array of tourism beneficiaries to offset visitor impacts and develop cost-sharing approaches. New research needs in this arena include:

1. Strategies for raising awareness and bolstering support for recreation and tourism on public lands to increase perceived value from host governments and private industry.
2. Exploration of public–private partnership models to encourage private investment in protected areas with a reasonable return.
3. Community investment in public lands opportunities and benefits of shared stewardship.

The lack of funding for parks and protected areas poses the biggest challenge to managing outdoor recreation and tourism to ensure that ecological systems are protected and conservation goals are achieved, while visitor experiences and community interactions are positive.

5.4 *Building institutional capacity*

Many parks and protected areas in both industrial and developing areas suffer from declining capacity, including human capacity (e.g., trained employees), financial capacity (e.g., budgets and flexibility in those budgets to respond to urgent needs), information capacity (e.g., scientific knowledge, access to data, standard tools and integrated frameworks), and physical capacity (e.g., vehicles, buildings, tools and machinery), which are all needed for safe and sustainable management and resilient systems. In some cases, tourism management may be undermined by other management goals. One respondent emphasized a focus on capacity needs:

> *A lack of funding poses the biggest challenge to managing or planning for outdoor recreation and tourism. This affects staffing, it affects planning and implementation efforts. The ability of land managers to maintain the current level of tourism on public land is highly threatened due to shrinking staff levels. So, development of new opportunities cannot happen.*

Persistent challenges in institutional capacity make it difficult for agencies to focus on long-term goals. The IUCN's Strategic Framework for Capacity Development recognizes the need for bolstering collective capacity for governance of public lands and protected areas to ensure long-term conservation goals are achieved (IUCN 2015). A four-fold process (a) encourages professionalism of protected area management, (b) bolsters relations with indigenous people and local communities, (c) develops strategic pathways for capacity development, and (d) uses indicators to measure effectiveness. Grassroots (bottom-up) approaches to capacity development may be considered that bolster training of middle managers and emphasize ongoing organizational learning (McCool et al. 2012). New approaches emphasize partnerships between conservation and tourism actors, such as the Global Sustainable Tourism Council, a strategic coalition of partners supporting global sustainable tourism (Bushell and Bricker 2017). Other examples of multi-stakeholder groups that add capacity to large-scale protected area management have emerged in states and provinces. Finally, some work has been initiated to examine how multiple resource agencies can examine their collective capacities for providing visitor experiences and looking for shared efficiencies. Research is needed in the following areas:

1. Approaches to build institutional capacity to govern protected area tourism.
2. Tools to expand human capacity by promoting professionalism, encouraging training opportunities, and building leaders.
3. Strategic approaches that manage recreation infrastructure and facilities at a landscape scale to encourage efficiencies across agencies.

More research on these approaches and their ability to build capacity for tourism management in protected areas would be of high value to land managers and communities.

Conclusion

Our public lands and protected areas provide a multitude of benefits to people who seek outdoor experiences and connections to nature, or who rely on natural resources for everyday needs. These natural spaces also conserve biodiversity and keep important ecosystems intact. There is growing recognition that our protected areas are not only a place to engage in outdoor leisure activities, but they are important also for individual and community health and resilience, livelihoods, and cultural survival. Being outdoors in nature has been shown to benefit people by improving fitness, health, cognition and stress reduction. Visitors to public lands provide economic benefits to host communities and can be an important source of employment. At the same time, demand for visiting protected areas and public lands is increasing and diversifying. Tourism patterns are changing on public lands, and planners and the scientific community are struggling to keep pace. Public agency capability to provide for sustainable tourism has been declining in many regions and land managers face challenges meeting demand, maintaining infrastructure and the integrity of natural and cultural resources, and providing quality visitor opportunities and services. Moreover, there is work to be done to support communities, where the economic benefits may not be extending to all and resulting socio-cultural changes suggest new challenges.

Based on our survey of 240 recreation and tourism professionals and scholars, we have identified five broad domains of research that require ongoing investment: social-ecological interactions, visitation trends, visitor experiences and benefits, community and culture, and governance, planning and sustainability monitoring. Within each domain, we have identified critical themes and outlined several priority areas for future research and capacity building (Figure 11.1).

Understanding tourism as a social-ecological system helps elucidate the dynamic interactions among institutions and processes and recognizes the multiplicity of voices with a stake in protected area management. Partnerships between scientific institutions and resource managers will ensure that knowledge production

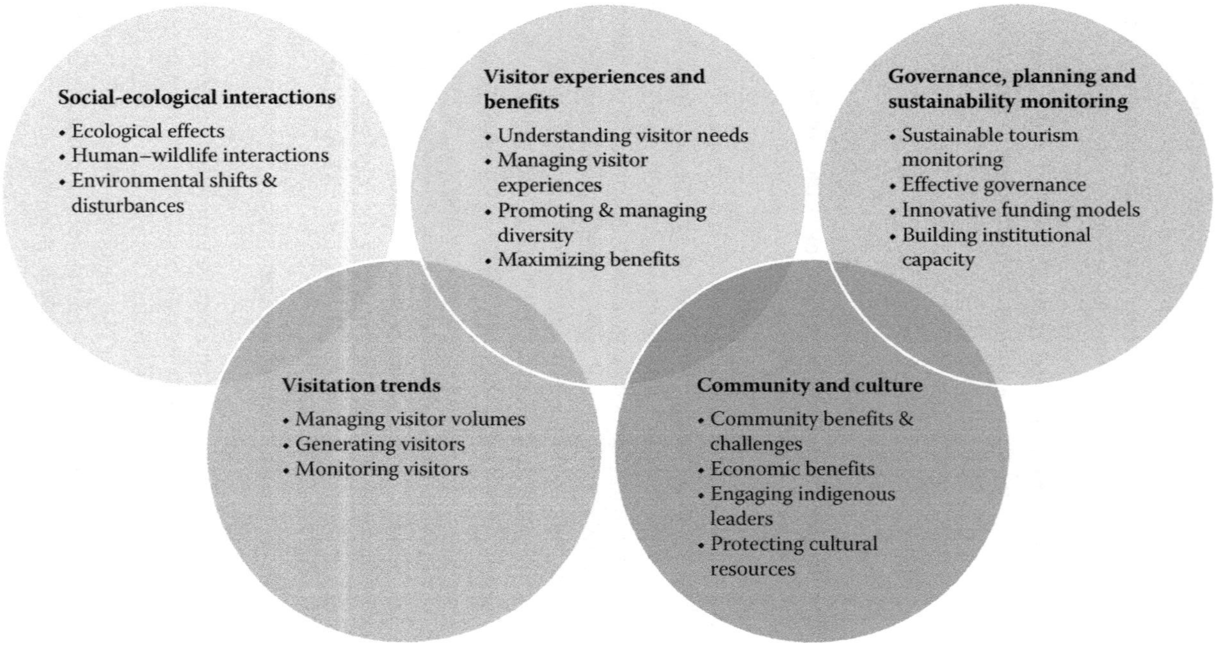

Figure 11.1 Research needs for sustainable protected area tourism

and exchange are multi-directional and mutually supportive. While needs for new studies are great, there also is a desire for knowledge syntheses and summaries that convene the latest research findings, trace the development of concepts, and provide an update on new methods for sustainable tourism management. Agencies that build capacity for knowledge production and exchange may benefit from institutionalizing best practices and case studies that demonstrate effective applications of scientific approaches and models for sustainable management. When ownership of science development, implementation and exchange is shared, the potential for positive impact is great.

References

Adams, W. M. and Hutton, J. (2007). "People, parks and poverty: political ecology and biodiversity conservation." *Conservation and Society*, **5**(2): 147–83.

Agrawal, A. and Redford, K. (2009). "Conservation and displacement: an overview." *Conservation and Society*, **7**(1): 1–10.

Andereck, K. L., Valentine, K. M., Knopf, R. C. and Vogt, C. A. (2005). "Residents' perceptions of community tourism impacts." *Annals of Tourism Research*, **32**(4): 1056–76.

Balmford, A., Beresford, J., Green, J., Naidoo, R., Walpole, M. and Manica, A. (2009). "A global perspective on trends in nature-based tourism." *PLoS Biology*, **7**(6): e1000144.

Balmford, A., Green, J. M., Anderson, M., Beresford, J., Huang, C., Naidoo, R., Walpole, M. and Manica, A. (2015). "Walk on the wild side: estimating the global magnitude of visits to protected areas." *PLoS Biology*, **13**(2): p.e1002074.

Bateman, P. W. and Flemming, P. A. (2017). "Are negative effects of tourist activities on wildlife over-reported? A review of assessment methods and empirical results." *Biological Conservation*, **211**: 10–19.

Becken, S. and Job, H. (2014). "Protected areas in an era of global–local change." *Journal of Sustainable Tourism*, **22**(4): 507–27.

Berno, T. and Bricker, K. (2001). "Sustainable tourism development: the long road from theory to practice." *International Journal of Economic Development*, **3**(3): 1–18.

Bricker, K. S., Black, R. and Cottrell, S. (eds.) (2013). *Sustainable Tourism and the Millennium Development Goals*. Burlington, MA: Jones & Bartlett.

Bricker, K. S. and Schultz, J. (2011). "Sustainable tourism in the USA: a comparative look at the global sustainable tourism criteria." *Tourism Recreation Research*, **36**(3): 215–29.

Brockington, D., Duffy, R. and Igoe, J. (2012). *Nature Unbound: Conservation, Capitalism and the Future of Protected Areas*. London: Routledge.

Bryce, R., Irvine, K. N., Church, A., Fish, R., Ranger, S. and Kenter, J. O. (2016). "Subjective wellbeing indicators for large-scale assessment of cultural ecosystem services." *Ecosystem Services*, **21**: 258–69.

Buckley, R. (2003). "Pay to play in parks: an Australian policy perspective on visitor fees in public protected areas." *Journal of Sustainable Tourism*, **11**(1): 56–73.

Buckley, R. (2009). "Parks and tourism." *PLoS Biology*, **7**(6): e1000143.

Buckley, R. (2012). "Sustainable tourism: research and reality." *Annals of Tourism Research*, **39**(2): 528–46.

Bushell, R. and Bricker, K. (2017). "Tourism in protected areas: developing meaningful standards." *Tourism and Hospitality Research*, **17**(1): 106–20.

Bushell, R. and Eagles, P. F. (eds.) (2006). *Tourism and Protected Areas: Benefits Beyond Boundaries: The Fifth IUCN World Parks Congress*. Wallingford: Cabi.

Butler, R. (2007). *Tourism and Indigenous Peoples: Issues and Implications*. London: Routledge.

Carr, A., Ruhanen, L. and Whitford, M. (2016). "Indigenous peoples and tourism: the challenges and opportunities for sustainable tourism." *Journal of Sustainable Tourism*, **24**(8–9): 1067–79.

CBD [Convention on Biological Diversity] (2011). "Strategic plan for 2011–2020 and the Aichi Targets." https://www.cbd.int/doc/strategic-plan/2011-2020/Aichi-Targets-EN.pdf.

Cerveny, L. (2008). *Nature and Tourists in the Last Frontier: Local Encounters with Global Tourism in Coastal Alaska*. Putnam Valley, NY: Cognizant Communication Corporation.

Cerveny, L. K., Blahna, D. J., Stern, M. J., Mortimer, M. J., Predmore, S. A. and Freeman, J. (2011). "The use of recreation planning tools in US Forest Service NEPA assessments." *Environmental Management*, **48**(3): 644–57.

Chandra, A. and Idrisova, A. (2011). "Convention on biological diversity: a review of national challenges and opportunities for implementation." *Biodiversity and Conservation*, **20**(14): 3295–316.

Choi, H. C. and Sirakaya, E. (2006). "Sustainability indicators for managing community tourism." *Tourism Management*, **27**: 1274–989.

Clark, R. N. and Stankey, G. H. (1979). "The recreation opportunity spectrum: a framework for planning, management, and research," p. 98. Gen. Tech. Rep. PNW-GTR-098. Portland, OR: US Department of Agriculture, Forest Service, Pacific Northwest Research Station.

Cole, D. N. (1981). "Managing ecological impacts at wilderness campsites and evaluation of techniques." *Journal of Forestry*, **79**(2): 86–9.

Cumming, G. S. and Allen, C. R. (2017). "Protected areas as social-ecological systems: perspectives from resilience and social-ecological systems theory." *Ecological Applications*. https://doi.org/10.1002/eap.1584.

Dawson, N., Martin, A. and Danielsen, F. (2018). "Assessing equity in protected area governance: approaches to promote just and effective conservation." *Conservation Letters*, **11**(2): 1–8.

Dhakal, N. P., Nelson, K. C. and Smith, J. D. (2011). "Resident well-being in conservation resettlement: the case of Padampur in the Royal Chitwan National Park, Nepal." *Society and Natural Resources*, **24**(6): 597–615.

Driver, B. L. and Brown, P. J. (1978). "The opportunity spectrum concept and behavior information in outdoor recreation resource supply inventories: a rationale." In H. G. Lund, V. J. LaBau, P. F. Folliott and D. W. Robinson (eds.), *Integrated Inventories of Renewable Natural Resources: Proceedings of the Workshop*. Gen. Tech. Report RM-55: 24–31. Ft. Collins, CO: USDA Forest Service, Rocky Mt. Forest and Range Exp. Station.

Dudley, N. (ed.) (2008). "Guidelines for applying protected area management categories." Gland: IUCN [International Union for Conservation of Nature]. https://cmsdata.iucn.org/downloads/guidelines_for_applying_protected_area_management_categories.pdf.

Eagles, P. F. (2002). "Trends in park tourism: economics, finance and management." *Journal of Sustainable Tourism*, **10**(2): 132–53.

Eagles, P. F. (2009). "Governance of recreation and tourism partnerships in parks and protected areas." *Journal of Sustainable Tourism*, **17**(2): 231–48.

Eagles, P. F. (2014). "Research priorities in park tourism." *Journal of Sustainable Tourism*, **22**(4): 528–49.

Eagles, P. F., McCool, S. F., Haynes, C. D. and Phillips, A. (2002). *Sustainable Tourism in Protected Areas: Guidelines for Planning and Management* (Vol. 8). Gland: IUCN.

Emerson, K., Nabatchi, T. and Balogh, S. (2012). "An integrative framework for collaborative governance." *Journal of Public Administration Research and Theory*, **22**(1): 1–29.

Espiner, S. and Becken, S. (2014). "Tourist towns on the edge: conceptualising vulnerability and resilience in a protected area tourism system." *Journal of Sustainable Tourism*, **22**(4): 646–65.

Fennell, D. and Weaver, D. (2005). "The ecotourism concept and tourism-conservation symbiosis." *Journal of Sustainable Tourism*, **13**(4): 373–90.

Ferraro, P. J. and Hanauer, M. M. (2011). "Protecting ecosystems and alleviating poverty with parks and reserves: 'Win-win' or tradeoffs?" *Environmental Resource Economics*, **48**: 269–86.

Fischer, D. M., Wood, S. A., White, E. M., Blahna, D. J., Lange, S., Weinberg, A., Tomco, M. and Lia, E. (2018). "Recreational use in dispersed public lands measured using social media data and on-site counts." *Journal of Environmental Management*, **222**: 465–74.

Fletcher, C., Pforr, C. and Brueckner, M. (2016). "Factors influencing indigenous engagement in tourism development: an international perspective." *Journal of Sustainable Tourism*, **24**(8–9): 1100–120.

Floyd, M. F. (2001). "Managing national parks in a multicultural society: searching for common ground." In *The George Wright Forum*, **18**(3): 41–51 (George Wright Society).

Frumkin, H., Bratman, G. N., Breslow, S. J., Cochran, B., Kahn Jr, P. H., Lawler, J. J., . . . and Wood, S. A. (2017). "Nature contact and human health: a research agenda." *Environmental Health Perspectives*, **125**(7): 075001.

Graefe, A., Vaske, J. and Kuss, F. (1984). "Social carrying capacity: an integration and synthesis of twenty years of research." *Leisure Sciences*, **8**: 275–95.

Graham, J., Amos, B. and Plumptre, T. W. (2003). "Governance principles for protected areas in the 21st century," pp. 1–2. Institute on Governance, Governance Principles for Protected Areas.

Greenwood, D. J. (1989). "Culture by the pound: an anthropological perspective on tourism as cultural commoditization." In V. L. Smith (ed.), *Hosts and Guests: The Anthropology of Tourism*, pp. 171–85. Philadelphia: University of Pennsylvania Press.

Hammit, W. E., Cole, D. N. and Monz, C. A. (2015). *Wildland Recreation: Ecology and Management*. 3rd edn. New York: John Wiley & Sons.

Hand, M. S., Smith, J. W., Peterson, D. L., Brunswick, N. A. and Brown, C. P. (2018). "Effects of climate change on outdoor recreation." In Jessica E. Halofsky, David L. Peterson, Joanne J. Ho, Natalie J. Little and Linda A. Joyce (eds.), *Climate Change Vulnerability and Adaptation in the Intermountain Region* [Part 2], pp. 316–38. Gen. Tech. Rep. RMRS-GTR-375. Fort Collins, CO: U.S. Department of Agriculture, Forest Service, Rocky Mountain Research Station.

Hutton, J., Adams, W. M. and Murombedzi, J. C. (2005). "Back to the barriers? Changing narratives in biodiversity conservation." *Forum for Development Studies*, **32**: 341–70.

IUCN [International Union for Conservation of Nature] (2015). "Strategic framework for capacity development in protected areas and other conserved territories 2015–2025." https://portals.iucn.org/library/sites/library/files/documents/Rep-2015-005.pdf.

IVUMC [Interagency Visitor Use Management Council] (2016). "Visitor Use Management Framework: a guide to providing sustainable outdoor recreation." Edition One. Available from https://visitoruse management.nps.gov/ (accessed November 10, 2018).

Job, H. and Paesler, F. (2013). "Links between nature-based tourism, protected areas, poverty alleviation and crises—the example of Wasini Island (Kenya)." *Journal of Outdoor Recreation and Tourism*, **1**: 18–28.

Jurowski, C. and Gursoy, D. (2004). "Distance effects on residents'attitudes toward tourism." *Annals of Tourism Research*, **31**(2): 296–312.

Karanth, K. K. and DeFries, R. (2011). "Nature-based tourism in Indian protected areas: new challenges for park management." *Conservation Letters*, **4**(2): 137–49.

Knight, R. L. and K. J. Gutzwiller (1995). *Wildlife and Recreationists: Coexistence through Management and Research*. Washington, DC: Island Press.

Larson, C. L., Reed, S. E., Merenlender, A. M. and Crooks, K. R. (2016). "Effects of recreation on animals revealed as widespread through a global systematic review." *PLoS ONE*, **11**(12): e0167259.

Lee, T. H. and Hsieh, H. P. (2016). "Indicators of sustainable tourism: a case study from Taiwan's wetland." *Ecological Indicators*, **67**: 779–87.

Leung, Y.-F., Spenceley, A., Hvenegaard, G. and Buckley, R. (eds.) (2018). *Tourism and Visitor Management in Protected Areas: Guidelines for Sustainability*. Best Practice Protected Area Guidelines Series, No. 27. Gland: IUCN.

Lischka, S. A., Teel, T. L., Johnson, H. E., Reed, S. E., Brekc, S., Don Carlos, A. and Crooks, K. R. (2018). "A conceptual model for the integration of social and ecological information to understand human-wildlife interactions." *Biological Conservation*, **225**: 80–87.

McCool, S. F. (2006). "Managing for visitor experience in protected areas: promising opportunities and fundamental challenges." *Parks*, **16**(2): 3–9.

McCool, S. F., Clark, R. N. and Stankey, G. H. (2007). "An assessment of frameworks useful for public land recreation planning." General Technical Report PNW-GTR-705. Portland, OR: US Department of Agriculture, Forest Service, Pacific Northwest Research Station. http://www.fs.fed.us/pnw/pubs/pnw_gtr705.pdf (accessed February 15, 2017).

McCool, S. F., Hsu, Y. C., Rocha, S. B., Sæþórsdóttir, A. D., Gardner, L. and Freimund, W. (2012). "Building the capability to manage tourism as support for the Aichi Target." *Parks*, **18**(2): 92.

McCool, S. F. and Spenceley, A. (2014). "Tourism and protected areas: a growing nexus of challenge and opportunity." *Koedoe*, **56**(2): Art. #1221, 2 pages.

McDonald, R. I., Forman, R. T., Kareiva, P., Neugarten, R., Salzer, D. and Fisher, J. (2009). "Urban effects, distance, and protected areas in an urbanizing world." *Landscape and Urban Planning*, **93**(1): 63–75.

McKercher, B. and Du Cros, H. (2002). *Cultural Tourism: The Partnership between Tourism and Cultural Heritage Management*. London: Routledge.

Marion, J. L., Leung, Y.-F., Eagleston, H. and Burroughs, K. (2017). "A review and synthesis of recreation ecology research findings on visitor impacts to wilderness and protected natural areas." *Journal of Forestry*, **114**(3): 352–62.

Mathieson, A. and Wall, G. (1982). *Tourism: Economic, Physical and Social Impacts*. Harlow: Longman.

Millennium Ecosystem Assessment (2005). *Millennium Ecosystem Assessment*. Washington, DC: New Island, p. 13.

Monz, C. A., Pickering, C. M. and Hadwen, W. L. (2013). "Recent advances in recreation ecology and the implications of different relationships between recreation use and ecological impacts." *Frontiers in Ecology*, **11**(8): 441–6.

Moore, R. L. and Driver, B. (2005). *Introduction to Outdoor Recreation: Providing and Managing Natural Resource Based Opportunities*. State College, PA: Venture Publishing.

Mosimane, A. W., McCool, S., Brown, P. and Ingrebretson, J. (2013). "Using mental models in the analysis of human–wildlife conflict from the perspective of a social-ecological system in Namibia." *Oryx*, **48**(1): 1–7.

Newsome, D., Moore, S. A. and Dowling, R. K. (2012). *Natural Area Tourism: Ecology, Impacts and Management* (Vol. 58). Cleveden: Channel View Publications.

Plummer, R. and Fennell, D. A. (2009). "Managing protected areas for sustainable tourism: prospects for adaptive co-management." *Journal of Sustainable Tourism*, **17**(2): 149–68.

Pohl, S. (2006). "Technology and the wilderness experience." *Environmental Ethics*, **28**: 147–63.

Pope, K. and Martin, S. R. (2011). "Visitor perceptions of technology, risk, and rescue in wilderness." *International Journal of Wilderness*, **15**(2): 19–26, 48.

Rebollo, J. F. V. and Baidal, J. A. I. (2003). "Measuring sustainability in a mass tourist destination: pressures, perceptions and policy responses in Torrevieja, Spain." *Journal of Sustainable Tourism*, **11**(2–3): 181–203.

Reinius, S. W. and Fredman, P. (2007). "Protected areas as attractions." *Annals of Tourism Research*, **34**(4): 839–54.

Saarinen, J. (2004). "'Destinations in change': the transformation process of tourist destinations." *Tourist Studies*, **4**(2): 161–79.

Sanchez, J. J., Baerenklau, K. and Gonzalez-Caban, A. (2016). "Valuing hypothetical wildfire impacts with a Kuhn-Tucker model of recreation demand." *Forest Policy and Economics*, **71**: 63–70.

Schägner, J. P., Arnberger, A., Eagles, P. F. J., Kajala, L., Leung, Y.-F., Spenceley, A., Deguignet, M., Gosal, A., Signorello, G., Engelbauer, M., Bertzky, B. and Engels, B. (2018). "Visitor numbers for protected and nature areas: a global data sharing initiative." *Proceedings of the 9th International Conference on Monitoring and Management of Visitors in Recreational and Protected Areas, August 28–31*. Bordeaux: MMV.

Scott, D., Amelung, B., Becken, S., Ceron, J. P., Dubois, G., Gössling, S., Peters, P. and Simpson, M. (2008). "Climate change and tourism: responding to global challenges." Caribbean Regional Sustainable Tourism Development Programme CTO Lot 3: Sustainable Tourism Policy Development. Madrid: World Tourism Organization, p. 230.

Sharpley, R. (2009). *Tourism Development and the Environment: Beyond Sustainability?* London: Routledge.

Sharpley, R. (2014). "Host perceptions of tourism: a review of the research." *Tourism Management*, **42**: 37–49.

Shultis, J. (2001). "Consuming nature: the uneasy relationship between technology, outdoor recreation and protected areas." *The George Wright Forum*, **18**(1): 56–66. George Wright Society.

Singh, R. K., Murty, H. R., Gupta, S. K. and Dikshit, A. K. (2009). "An overview of sustainability assessment methodologies." *Ecological Indicators*, **9**(2): 189–212.

Smith, V. L. and Brent, M. (2001). *Hosts and Guests Revisited: Tourism Issues of the 21st Century*. Putnam Valley, NY: Cognizant Communication Corporation.

Spence, M. D. (1999). *Dispossessing the Wilderness: Indian Removal and the Making of the National Parks*. Oxford: Oxford University Press.

Spenceley, A. (ed.) (2008). *Responsible Tourism: Critical Issues for Conservation and Development*. London: Earthscan.

Stankey, G. H., Cole, D. N., Lucas, R. C., Petersen, M. E. and Frissell, S. S. (1985). "The Limit of Acceptable Change (LAC) System for Wilderness Planning." General Technical Report INT-176. Ogden, UT: USDA Forest Service, Intermountain Research Station.

Strickland-Munro, J. K., Allison, H. E. and Moore, S. A. (2010). "Using resilience concepts to investigate the impacts of protected area tourism on communities." *Annals of Tourism Research*, **37**(2): 499–519.

Stronza, A. (2007). "The economic promise of ecotourism for conservation." *Journal of Ecotourism*, **6**(3): 210–30.

Tablado, Z. and Jenni, L. (2017). "Determinants of uncertainty in wildlife responses to human disturbance." *Biological Reviews*, **92**: 216–33.

Tanguay, G. A., Rajaonson, J. and Therrien, M. C. (2013). "Sustainable tourism indicators: selection criteria for policy implementation and scientific recognition." *Journal of Sustainable Tourism*, **21**(6): 862–79.

Timothy, D. J. and Nyaupane, G. P. (2009). "Protecting the past: challenges and opportunities." In Timothy and Nyaupane (eds.), *Cultural Heritage and Tourism in the Developing World: A Regional Perspective*, pp. 34–55. Abingdon: Routledge.

Twohig-Bennett, C., and Jones, A. (2018). "The health benefits of the great outdoors: a systematic review and meta-analysis of greenspace exposure and health outcomes." *Environmental Research*, **166**: 628–37.

UNEP and IUCN [United Nations Environment Programme, World Conservation Monitoring Center and International Union for Conservation of Nature] (2016). "Protected Planet Report 2016." Cambridge and Gland: UNEP–WCMC and IUCN. https://www.protectedplanet.net/c/protected-planet-report-2016.

UNWTO [United Nations World Tourism Organization] (2005). "Making tourism more sustainable – a guide for policy makers," pp. 11–12. http://sdt.unwto.org/content/about-us-5.

UNWTO [United Nations World Tourism Organization] (2018). "World tourism barometer." http://media.unwto.org/press-release/2018-01-15/2017-international-tourism-results-highest-seven-years.

Walden-Schreiner, C., Leung, Y.-F. and Tateosian, L. (2018). "Digital footprints: incorporating crowdsourced geographic information for protected area management." *Applied Geography*, **90**: 44–54.

Watson, J. E., Dudley, N., Segan, D. B. and Hockings, M. (2014). "The performance and potential of protected areas." *Nature*, **515**(7525): 67.

Weiermair, K. and Mathies, C. (2004). *The Tourism and Leisure Industry: Shaping the Future*. Hove: Psychology Press.

West, P., Igoe, J. and Brockington, D. (2006). "Parks and peoples: the social impact of protected areas." *Annual Review of Anthropology*, **35**: 251–77.

White, E. M. (2016). "Federal outdoor recreation trends." General Technical Report PNW, p. 945. US Department of Agriculture, Forest Service, Pacific Northwest Research Station.

Whitelaw, P. A., King, B. E. and Tolkach, D. (2014). "Protected areas, conservation and tourism–financing the sustainable dream." *Journal of Sustainable Tourism*, **22**(4): 584–603.

Wilkins, E. J., Smith, J. W. and Keane, R. (2018). "Social media communication preferences of national park visitors." *Applied Environmental Education & Communication.* doi: 10.1080/1533015X.2018.1486247.

Wood, S. A., Guerry, A. D., Silver, J. M. and Lacayo, M. (2013). "Using social media to quantify nature-based tourism and recreation." *Scientific Reports,* **3**: 2976.

WTTC [World Travel and Tourism Council] (2018). "Travel and tourism economic impact, 2018." https://www.wttc.org/-/media/files/reports/economic-impact-research/regions-2018/world2018.pdf.

Wunder, S. (2000). "Ecotourism and economic incentives—an empirical approach." *Ecological Economics,* **32**: 465–79.

12 Monitoring river visitor use in a complex social-ecological system

Jennifer Thomsen, Iree Wheeler, Douglas Dalenberg and Colter Pence

Introduction

Complexity of the river system

River systems encompass complex social and ecological components that transcend political and jurisdictional boundaries (Chaffin, Gosnell & Cosens 2014). Rivers attract diverse types of recreation (referred to in this chapter as tourism although this does include local and non-local use) such as fishing, rafting, kayaking, paddle boarding and swimming. Some of this tourism occurs on the river, while other forms of recreation, like fishing, can also occur along the riverbanks. Additionally, rivers attract local users as well as tourists from around the country and the world. Beyond recreation, rivers provide many ecosystem services (Auerbach et al. 2014; Wenger et al. 2011). Water is increasingly a resource in short supply and high demand leading to conflicts worldwide of how to prioritize use and management of the precious resource (Postel 2014).

The ecological components of the river system offer additional complexity as the river influences and connects the landscape. The system is impacted by a variety of factors such as climate change and invasive species (Hirabayashi et al. 2013). Management requires resilience thinking to respond to changes and shifts in the system as well as uncertainty of future changes (Folke et al. 2010). Monitoring the river system is critical to truly understand when and how change occurs as well as how to respond adaptively.

Importance of monitoring river systems

Monitoring is a crucial aspect in a planning and management process and is important to assess whether or not desired conditions are being met, evaluate a new management action, or identify when changes occur in the system and adaptive management is needed (Allen et al. 2011). Monitoring visitor use and tourism on rivers requires diverse strategies with unique tools that can provide insights employing a variety of indicators and thresholds measuring whether desired ecological and social conditions are being met. Indicators are "specific resource or experiential attributes that can be measured to track changes in conditions that progress toward achieving and maintaining desired conditions" (IVUMC 2016, p. 38). Indicators for rivers related to experience can include the number of boat encounters or the time it takes to put-in or take-out on the river. Ecological indicators for rivers can include erosion percentage of the riverbanks, compaction of soil at put-in or take-outs, water quality levels or species population health. It is important when selecting indicators that they are sensitive to change, reasonable, reliable and important to the social-ecological system (IVUMC 2016).

Thresholds are "minimally acceptable conditions associated with each indicator" (IVUMC 2016, p. 38). It is important to note that thresholds should be informed by data and science but establishing thresholds can also be influenced by values. Monitoring indicators offer an opportunity to implement management actions prior to reaching the threshold mark. However, there are many types of data to be collected and diverse indicators to be monitored adding to the complexity of river management.

There are several examples that illustrate the importance of monitoring use and tourism on rivers. For instance, the Colorado River in the U.S. is a complex system which is in high demand for recreational services, while also acting as an economic driver of the region and serving as a major water source in an area with limited water supply (Taber 2012). With the competing demands for the Colorado River, the Colorado River Management Plan Environmental Impact Statement determined primary factors that influence user capacities to maximize user experience and minimize ecological impacts. These factors include physical aspects (i.e., number, size, distribution and vulnerability of camping areas along river system), resource aspects (i.e., types and condition of natural and cultural resources along river system), and social aspects (i.e., contacts per day on the river and at attractions, number of trips in the canyon at one time, number of people in the canyon at one time, group size, trip length and launch patterns) (National Park Service 2006). The Colorado River serves as an example of the complex social, ecological and economic aspects of a river system.

Similar to the Colorado River, the Flathead River System, located in Montana, is embedded within a multifaceted social-ecological system. This chapter aims to address challenges in managing sustainable recreation and tourism use of a river resource. Using the three forks of the Flathead River System as an example, the chapter discusses various strategies of monitoring rivers for visitor use on and off the river, as well as public versus private use. The chapter further identifies strategies for analyzing and interpreting data, concluding with key lessons and recommendations for research topics to stimulate future growth in monitoring and managing a river system for sustainable tourism.

The complexities of the Flathead River System

Located in Northwest Montana in the U.S., the Flathead River System is 219 miles of free-flowing river, comprising three forks: the North Fork, the Middle Fork, and the South Fork (Figure 12.1). The Flathead River includes "the North Fork from

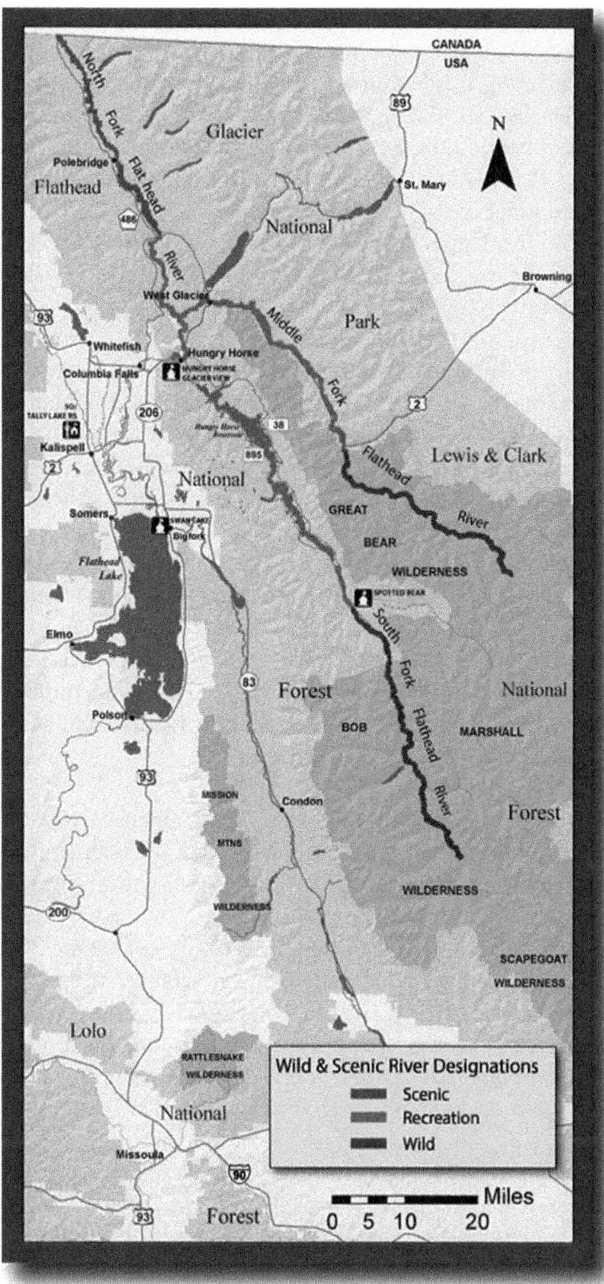

Source: U.S. Forest Service (2015): available at https://www.fs.usda.gov/detailfull/flathead/home/?cid=fseprd573051&width=full.

Figure 12.1 Map of the Flathead River System and the Wild and Scenic River designations for sections of the system

the Canadian border to its confluence with the Middle Fork, the entire Middle Fork, and the South Fork from its headwaters to the Hungry Horse Reservoir" (U.S. Forest Service 1980). While much of the Flathead River System flows through United States Forest Service (USFS) administered lands, most of the North Fork and large portions of the Middle Fork comprise the western and southern borders of Glacier National Park. Due to the shared shores of the river system, the North Fork of the Flathead River System is jointly managed by the USFS and the National Park Service (NPS). The Upper Middle Fork and the South Fork of the Flathead River flow through the rugged Bob Marshall and Great Bear Wilderness areas, which are managed by the USFS.

In 1976, the Flathead River was designated as a Wild and Scenic River (WSR), under the Wild and Scenic River Act. Congress established the National Wild and Scenic Rivers System in 1968 "to preserve certain rivers with outstanding nature, cultural, and recreational values in a free-flowing condition for the enjoyment of present and future generations." Wild and Scenic Rivers offer a variety of ecological benefits including conservation of flora and fauna, filtering water, and sustaining ecosystem functioning (American Rivers 2017). Additionally, the designation supports ecosystem services such as clean water, reduces the impacts of flooding, and offers exceptional recreational opportunities (American Rivers 2017). The WSR designation of the Flathead River System positions managers in a unique and challenging situation to maintain and enhance the "outstanding remarkable values" noted in the Act, which facilitated the designation in the first place, while also providing sustainable recreational opportunities.

As a WSR, the Flathead River System requires a Comprehensive River Management Plan (CRMP) that identifies the outstandingly remarkable values (ORVs) and the goals and desired conditions for protecting those values within the system:

> In order to be considered as outstandingly remarkable, a river-related value must be a unique, rare, or exemplary feature that is significant at a comparative regional or national scale (region of comparison). Values are scenic, recreational, geological, fish related, wildlife related, historic, cultural, botanical, hydrological, paleontological, scientific, or other values. While the spectrum of resources that may be considered is broad, all values should be directly river related. (BLM Spokane District Office 2010, p. 1)

These outstanding remarkable values inform the classification of river sections as either Wild, Scenic, or Recreation Rivers (Figure 12.1). For the Flathead River, the upper North Fork is classified as Scenic, which is "a river or segment of a river that is free of impoundments, with shorelines or watersheds still largely primitive and shorelines largely undeveloped, but accessible in places by roads" (USFS & NPS 2013, p. 1). The lower North Fork, lower Middle Fork and lower South Fork are classified as Recreational, which is "a river or segment of a river that is readily accessible by road or railroad, that may have some development along its shorelines, and that may have undergone some impoundment or diversion in the past" (USFS & NPS 2013, p. 1). The upper Middle Fork, upper South Fork and the central South

Fork extend through Bob Marshall Wilderness and the Great Bear Wildernesses. These sections are classified as Wild, which is "a river or segment of a river that is free of impoundments and generally inaccessible except by trail, with watersheds or shorelines essentially primitive and waters unpolluted. These represent vestiges of primitive America" (USFS & NPS 2013, p. 1). The associated classifications of the various segments of the Flathead River illustrate the complexity in managing the system to maintain different ORVs.

The Flathead River system exists in the context of rapid social and environmental change. Visitation to Glacier National Park has experienced exponential growth reaching three million growth visitors in 2017 (National Park Service 2018). The population of the Flathead Valley surrounding the river is also increasing at a rapid rate. For example, Flathead County grew by 10 percent between 2010 and 2017 (U.S. Census Bureau 2017). More people in the area can put more pressure on the river system. In 2017, the number of guided trips by tour operators, typically referred to as outfitters, increased by 33 percent between 2013 and 2017 on the North and Middle Fork sections of the river (U.S. Forest Service 2018). Additionally, there are new types of users in the corridor. Watercraft, such as pack rafts, offer an affordable and lightweight option for river users to carry a deflated raft on their back like a pack and then inflate the raft once they hike to the river. As a result of this growing trend, many recreationists do not need a guide to help them navigate rapids on the river and once inaccessible sites for river access are now accessible (Great Falls Tribune 2014). As a result, the Flathead River is experiencing more groups on the shores of the river for fishing and camping in addition to on-river use and extended seasons. This could affect peoples' perceptions of crowding depending on whether they are on the river or on shore. Increasing visitation may also increase impacts along riverbanks and put-in and take-out areas along the river system.

Strategies for monitoring river use

Surveying users to inform indicators and standards

Surveying river users and tourists is an invaluable tool to help establish and update indicators and thresholds. The 1986 Flathead Wild and Scenic River Recreation Management Direction, a follow-up document to the 1980 Management Plan, aims to "resolve the allocation and rationing issue using new information based in a large part on the results of a two-year (1980–81) study of river floaters" (U.S. Forest Service 1986, p. 1). The 1986 Direction document was based on the recreation opportunity spectrum (ROS) (Clark and Stankey 1979) and the limits of acceptable change (LAC) (Stankey et al. 1985) framework to address the diverse sections of the Flathead River System and as a strategy to balance visitor use with preservation of ORVs (U.S. Forest Service 1986). The main indicators include encounters per day with other groups floating the river, encounters per day with other groups on shore, campsite conditions, experience quality, occurrence of litter, kilometers of shoreline with permanent human-made modifications in foreground, and mechanical sounds from watercraft, defacement of natural features, changes in use patterns, and congestion at launch sites (U.S. Forest Service 1986). The establishment of indicators and standards based on resource and user data provides managers with clear and measurable metrics to assess changes in the resource or visitor experience that can be monitored over time.

An initial survey can help inform standards, but conducting periodic follow-up surveys is important in assessing whether experiences and expectations are changing among river users and tourists. In 2012, the Forest Service in conjunction with Glacier National Park, surveyed users of all three forks of the Flathead River corridor, including floaters, backpackers and horseback groups. The surveys were administered by rangers at a variety of sites along the river and were also given to outfitters to provide to their clients at the end of their trip. The survey collected information about trip characteristics and specific details from those who floated the river and from those who camped along the river. The survey asked questions on a variety of topics: (1) encounters and their impacts on respondents' experiences; (2) changing site conditions for repeat visitors; and (3) problems users might have encountered and potential management actions. The survey provides an outstanding snapshot of the quality of the resource and the visitor experience of using the resource along with information about how the resource is changing from the perspective of repeat visitors, along with reactions to possible management alternatives. While this survey has not yet informed substantial changes to management, the survey is a tool to inform managers on trends that can be assessed across indicators and thresholds.

Developing a strategy for on-river monitoring

Monitoring how many people are using river resources is important for guiding management decisions and maintaining integrity of resources. Different groups and agencies have approached monitoring river users in diverse ways. These approaches help create a larger comprehensive understanding of visitor use for mangers. The array of monitoring tools includes the collection of river-use data from on-river and on-shore perspectives.

In the case of the Flathead River System, the USFS has approached monitoring of river recreation using on-river monitoring of visitor encounters by USFS river rangers. This monitoring approach provides several key insights to understanding river recreation use. The information monitored by the USFS river rangers is aligned with the indicators established in the1986 Flathead Wild and Scenic River Recreation Management Direction (i.e., encounters with float parties, encounters with shore parties, mechanical sounds heard, and occurrences of litter or debris). Within the category of "on river encounters with float parties," the number of craft, type of craft, party size, whether the trip was guided or not, and whether they were close to an access site are recorded. The category of "encounters with shore parties" distinguishes the type of party based on their recreational activity (U.S. Forest Service 2018).

On-river monitoring is beneficial in informing management decisions and assessing whether desired conditions are being met. This monitoring tool also allows for interactions between users and river rangers which provide rangers with a perspective of the river users' experience. Additionally, the USFS has been conducting on-river visitor monitoring using a standardized monitoring form since 1997. The monitoring process has allowed for a longitudinal perspective of how visitor use has changed and developed (U.S. Forest Service 2018). The on-river monitoring conducted for the Flathead River System is utilized by other river managers and is also commonly used in wilderness monitoring for trail encounters.

Notwithstanding all the benefits of on-river monitoring, there are some limitations to using this approach. Since monitoring began in 1997, USFS river ranger staff levels have decreased leading to fewer river patrols per season while the number of recreation users has increased. For example, in 1998 the USFS conducted 68 patrols, but by 2015 the number had dropped to 43 (U.S. Forest Service 2018). There have also been changes in the times of day and week when river rangers patrol the river. They used to float a standard 8:00 a.m. start time Monday–Friday while river users often float on weekends and later in the day. River rangers now float on weekends and begin later in the morning to get a more representative sample (U.S. Forest Service 2018).

Using this monitoring tool on its own only provides a snapshot of daily use impacts on a stretch of river for a limited sample of days. When river rangers are floating, they only see users who are on the river in their direct proximity at that period of time and day. This type of data collection leaves a gap in understanding what a shore party or landowner who remains stationary on the shore might see in a day and for understanding actual counts of comprehensive river use throughout the river season despite day, time and stretch of river.

Additionally, it is challenging to update and maintain the database due to financial costs. As a result, the USFS is limited in the types of analysis and visual displays of data they can produce. For example, geofencing, the practice of using global positioning (GPS) or radio frequency identification (RFID) to define a geographic boundary, can act as a tool to create a virtual fence that can trigger text messaging, notifications, or questions for tourists and river users within a specified area (Chamberlain 2016).

Developing a strategy for on-shore river monitoring

On-shore monitoring techniques offer a different and complementary tool to understand river use. On the Flathead River System, the University of Montana is collecting actual counts versus sample data regarding how many boats, parties and people come down given stretches of the Flathead River every day. This data is collected using stationary outdoor Plotwatcher Pro game cameras installed on-shore along different popular river sections. These cameras take a photo every three to five minutes, depending on time of year and water speed, the goal being to capture each boat that passes without getting too many duplicated images of the same groups. This is important because all of the image files are played using outdoor Gamefinder software. This program displays all of the images in a stop-motion film, and having fewer images to play not only shortens watching time but allows the cameras to be left in the field for longer periods of time. The cameras are set to turn on at 7:30 a.m. in the beginning of the season and turn off at 9:30 p.m. These times are adjusted as the days get shorter, ending the season turning on at 8:00 a.m. and turning off at 8:00 p.m. The cameras are equipped with 32 GB SD cards which are swapped out every two to three weeks, depending on how frequently images are being taken.

The locations of the river monitoring sites were selected through consultation with the USFS, NPS and state resource managers. This method of site selection allowed mangers with different perspectives to prioritize certain locations based on need for monitoring data. In 2018, site specific visitation and use data was collected for 12 stretches of river (limited to this number due to constraints on the researching team of servicing equipment and processing data for more sites).

During the analysis stage, the camera footage is uploaded to a computer and every image is viewed manually (Figure 12.2). In each image, data is recorded for how many watercraft and float parties. Additionally, the date, time and number of

Source: University of Montana.

Figure 12.2 Example of camera data footage during analysis

people and boats are recorded as well as whether the craft is private or commercial. Determining whether a craft is private or commercial can be difficult due to visibility in the images and the similar colors of commercial and private crafts, so this number is crosschecked with data provided by outfitters/guides to the USFS to ensure accuracy.

The most significant benefit of using this on-shore monitoring approach is the census of river visitation and use. Every boat and person that passes through a specific site is recorded providing managers with an accurate understanding of actual use on a stretch by day, month and season, and serves as a foundation for an effective sampling approach in the future. Additionally, by capturing every craft that passes a given fixed location, this monitoring strategy provides a unique perspective from the landowner or on-shore recreational user. Another benefit of using monitoring equipment at a variety of locations is that this approach can encompass more river miles than the USFS on-river monitoring method and capture data throughout the day and season on all stretches of the river. The data also allows for managers to compare use across different stretches, and forks, of the Flathead River System.

Similar video-based recreation monitoring has been used successfully in Astoria, Oregon, to monitor ocean recreational fishing efforts originating in the Columbia River (Edwards & Schindler 2017) and to monitor recreational boats departing from Newport, Oregon (Ames & Schindler 2009). In these studies, web-based video cameras were used, although this method is not feasible for the Flathead River System due to the remote locations with no cell or internet access. Similar video footage to monitor angler use has been employed on lakes in British Columbia (Greenberg & Godin 2015) and in the Gulf of Mexico Red Snapper Fishery (Powers & Anson 2016). King County in Washington offers an example of where a variety of methods were used to comprehensively monitor river recreation use on several rivers within the county. Monitoring tools employed included aerial surveys, remote camera observations, field observations and in-person interviews (Herrera Environmental Consultants 2014).

There are limitations of using this monitoring tool alone. For example, there is a lack of longitudinal data using the camera equipment versus several decades of data captured by the USFS on-river monitoring. Additionally the information collected is not as in-depth as that collected in on-river monitoring. Because of the wide range of locations being monitored, it has only been feasible to record party size, number of watercraft, and number of people in each craft. This data does not provide insight into recreation activities or watercraft type. In addition to the upkeep of field equipment, the processing of video footage for the Flathead River on-shore monitoring data takes approximately ten hours per two-week period. For the 2018 river monitoring year, the analysis process resulted in approximately 840 hours to watch the 14 weeks of data from 12 sites. Lastly, the cameras do not offer an on-river perspective of the tourism and user experience, limit interaction between USFS staff and river users, and reduce monitoring for other impacts to resources.

Developing a strategy for tour operator/outfitter use monitoring

Outfitter (used as analogous to tour operator) use on the Flathead River is a popular activity especially for tourists visiting the area. Outfitter trips are run each year with five outfitters who have permits for a certain number of service days on specific forks of the river. The permit system is for a ten-year period and at the end of the term, a process required of the National Environmental Policy Act (NEPA) is conducted to issue new permits. During this process, the USFS evaluates the potential impacts to natural and social resources which inform use level allocation. Every five years, the amount of use is evaluated to determine if the outfitter is sufficiently using their allocations. Outfitter permits up for renewal are reissued to the holders unless there are performance issues that remain unresolved or if they do not apply for them to be reissued.

Due to the different classifications for each section of the Flathead River, outfitter use limits are based on interpretation of the 1980 River Management Plan and the 1986 Recreation Direction document; therefore, the number of permits and use levels have remained mostly static from the original allocations established in the late 1970s to early 1980s. Most of the trips are day-use, but outfitters also offer multi-day trips. Outfitters and guides who operate on the Flathead River are required to report their trips each season to the Forest Service. The USFS has been collecting the commercial use data since permits were allocated and in some permit cases, since 1977. The information gathered from these outfitters and guides through conversations and seasonal reports helps to provide depth to the information gathered from on-river and on-shore monitoring efforts. The main data collected includes the date of the trip, stretch of the river, launch time, type of trip, number of crafts per trip and number of clients per trip. Through discussions with the outfitters, other challenges have emerged including limitations in access to sites and competition for popular fishing sites (focus group with outfitters 2018). As pack rafts and mobility of travelers and gear evolve, there may be additional opportunities for commercial users in the future.

It is important to note that while commercial use has been operating on a regulated permit system, private use trips have not required permits. As the management plan development process for the Flathead River proceeds, the various forms of data collection are proving critical to informing whether different management actions, including the possibility of a permit system for private use, are options.

Developing a strategy for analyzing, synthesizing and interpreting data

The information that river managers collect on the health and use of the river is invaluable. Managers need resources and personnel to produce summaries including graphs and tables to help interpret the data. In some cases, commercial software such as TrafX for infrared and magnetic counters or TraxPro for tube counters can be very helpful for organizing and analyzing data.[1] Careful documentation of the

1 See https://www.trafx.net and https://www.jamartech.com/traxpro.html, respectively.

procedures and standards used to collect and record the data requires organizational skills and attention to detail. Checking the data for errors or unusual observations requires examining the data in summary and in some cases examining individual observations (Long 2009). Data cleaning requires a myriad of decisions from simple exclusions of data that were recorded in error due to equipment malfunctions, to difficult choices involving special circumstances noted by observers. Careful thought should be given to how the data is to be stored, backed up and managed both for internal use and external constitutes. Data stored in a spreadsheet can be analyzed within the spreadsheet and with statistical programs.

Synthesis of on-river and on-shore monitoring approaches

One challenge that even an advanced analyst faces is that river monitoring data originates from a number of different sources, each with its own characteristics and time dimensions. For example, consider on-river monitoring in contrast with an on-shore monitoring census generated by cameras. The on-river floats each have a river section, a specific starting and ending time, and specific encounter times along the float. It would be valuable for river managers to know how many rafts passed the camera that day while they were floating on the camera's section of the river. This would require merging the camera data with the float data based on the date. Once the data set is constructed, the two methods' counts can be compared and contrasted.

When combined, on-river and on-shore monitoring approaches provide insight to managers on which sections of river get the most use, what types of activities visitors are participating in, and what a visitor might see from both an on-shore perspective as well as while floating down the river. Figures 12.3 and 12.4 offer an example of how on-river and on-shore monitoring can be used together to illustrate a more complete story of visitor use. Figure 12.3 displays the average number of float parties on the Upper North Fork between 1997 and 2017. This data shows that in 2017, the USFS river rangers encountered an average of two parties while on the Upper North Fork. Figure 12.4 shows a comparison of use from the on-shore perspective from Wurtz airstrip between 2017 and 2018. This stretch of river is the northernmost and most remote stretch of the North Fork within the United States. This monitoring site captures floaters who have launched at the border between the United States and Canada, which is 58 miles from the nearest town, reached mainly by a dirt road. From this on-shore data, almost 80 watercraft passed by the monitoring site on high-use days in July 2017, though the average for July 2017 was 21 watercraft.

The differences in these monitoring numbers demonstrates the diverse perspectives of river use at this location throughout the season. While use is generally low, there are instances (i.e., weekends in July) when use climbs significantly in this area. By combining the on-shore and on-river monitoring, managers have a more comprehensive understanding of the pressures facing the river.

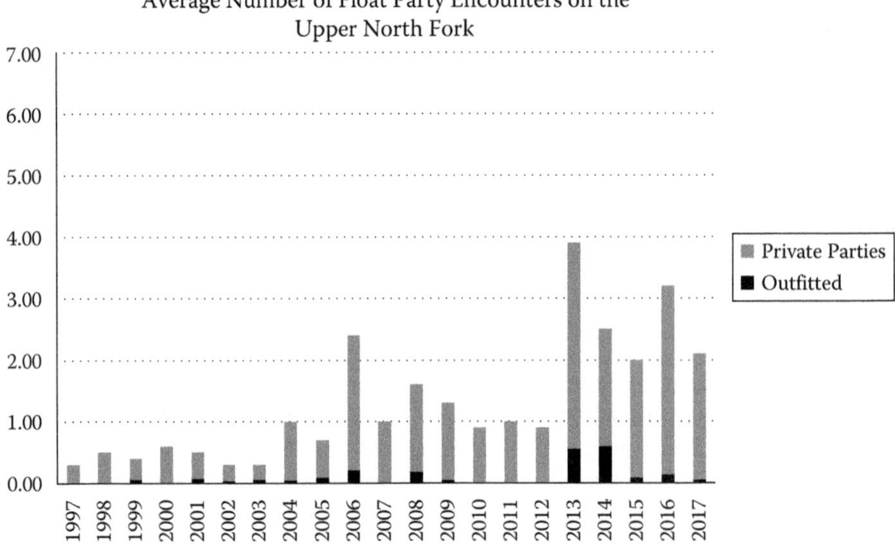

Figure 12.3 Example of on-river USFS monitoring

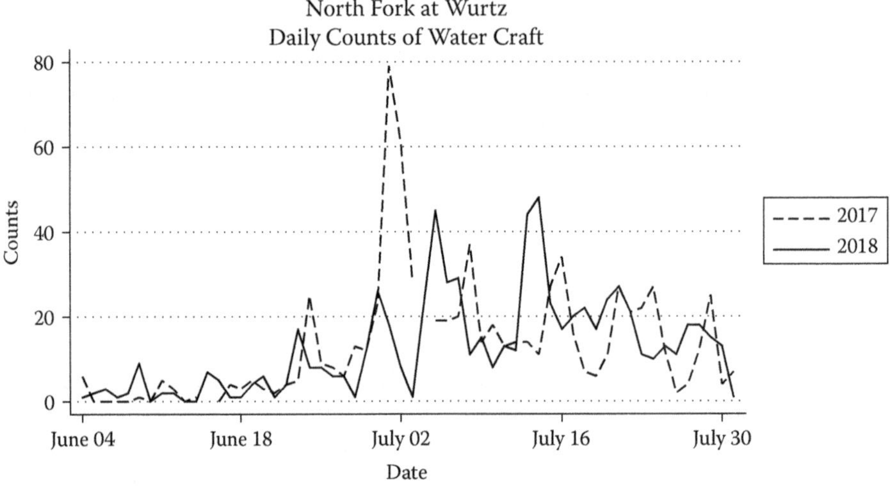

Figure 12.4 Example of on-shore monitoring

The census data would also be valuable to a statistician designing a sampling schedule to make the float data more representative of the population. For example, in 2017 on the North Fork, 38 percent of all float craft passed the camera on a weekend. Therefore on-river float monitoring could be weighted to test the stated standards or monitoring could be adjusted to have approximately 38 percent of

Table 12.1 Monitoring versus camera counts: Lower Middle Fork 2017

Trip	Party count (monitoring float)	Party count (camera daily total)	Craft count (monitoring float)	Craft count (camera daily total)	People count (monitoring float)	People count (camera daily total)
1	0	20	0	34	0	121
2	10	88	14	153	32	418
3	6	45	11	115	43	267
4	4	51	9	93	50	284
5	2	19	7	37	14	121
6	1	31	1	58	4	222
7	0	24	0	58	0	213
8	8	47	9	139	30	332
9	2	17	3	43	6	86
Correlation	0.88		0.83		0.66	

Note: Correlation measures the strength of a linear relationship with high positive correlations representing a strong positive linear relationship. In this case the high correlations indicate that if a monitoring float value is high relative to its mean, then the camera data tends to be high relative to its mean.

the monitoring floats on the weekend. This would allow the sample which is constructed with on-river float monitoring to better reflect the true population which is being captured by the camera.

In 2017, nine USFS monitoring floats on the Lower Middle Fork were on a section of the river that passed a particular on-shore camera location. Table 12.1 compares the float data for each river ranger trip along that section of the Lower Middle Fork to the daily totals that the camera captured on that same section. Although the numbers appear quite different, the correlations between the on-river float data and the camera data are quite high for this section of the river. For parties, the correlations between float monitored parties and camera counts for the day is 0.88 and for number of craft it is 0.83. This means that when the camera recorded a busy day on-shore, the monitoring on-river float recorded a busy day (i.e., trips 2 and 3). Similarly, when the camera recorded a quiet day, the monitoring float recorded a quiet day (i.e., trips 1, 5 and 9). Camera counts can be used to ensure that the on-river monitoring plan is capturing an accurate picture of visitor use for different sections of the river.

Camera data is also valuable in identifying potential monitoring issues. For example, the correlation between the number of people monitored on-river with the USFS and on-shore with the camera data is 0.66, which results from a few large parties that may not be captured as well by the on-river monitoring. Adding automobile traffic counts near popular access sites, coupled with the camera data or river monitoring data, would help establish whether car counts are correlated with river traffic and could assist with monitoring. In addition to cameras along the river corridor, this project includes monitoring traffic in launch sites using magnetic road counters. These counters are programmed to count every time a large metal object passes by. The counters are calibrated using cameras to ensure accuracy. Launch sites have been expressed as a main source of crowding within the recreational-use experience. The USFS rangers also monitor wait time at launch sites when floating the river. This launch-site traffic data will add to the understanding of congestion at these locations.

Regardless of the data source, constructing well-designed visuals to aid in analysis and dissemination assist in interpreting the data. In particular, choosing the appropriate visual by thinking about what data needs to be shown is vital (Tufte 1983). Steele, Chandler & Reddy (2016) lay out principles of data visualization with examples of both good and bad graphics created with R statistical program. For example, Figure 12.5 displays the daily distribution of trips using a histogram (recommended) and a pie chart (not recommended). Data visualization experts caution against using pie charts since people are not good at judging areas in them. Brief written summaries along with appropriate graphs, often focused on specific questions, can be helpful in putting the big picture together and identifying questions that the current data cannot answer.

Presenting the data analysis and getting feedback, although time consuming, can pay large dividends. River managers also need to remember that observational

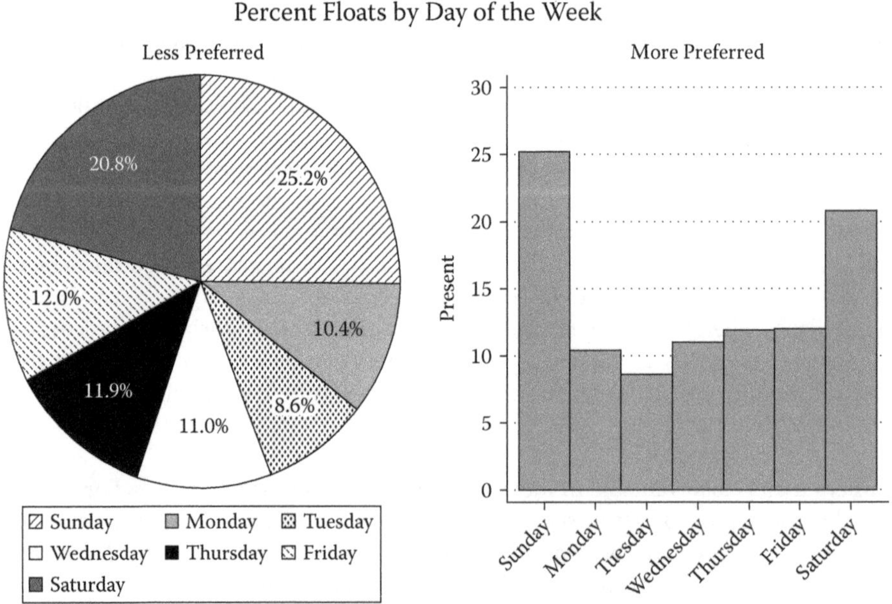

Figure 12.5 Examples of options for effectively displaying and interpreting data

data provides valuable descriptions but cannot establish causation, especially in a complex river system. While the data is a vital aid, it cannot substitute for careful thought and a deep study of the river system. Dissemination beyond local internal sources also will raise new questions and issues and generate new insights which will contribute to that critical thinking.

For managers, the major challenge is finding resources and individuals with the time and software skills to maintain and analyze a comprehensive dataset. Developing a sampling protocol with indicators and thresholds for both the ecological resource and user experience may require extra effort and consultation with statisticians skilled in applied sampling in order to ensure that the sample data reflects the true underlying population data. Processing and managing the data can be a significant challenge that can be overcome by assigning a data manager and documenting processes for data collection and management. However, the most difficult challenge may be keeping a consistent and well-documented archive of data over time. While analyses of short time periods of data are undoubtedly valuable, in order to evaluate a river systems' resilience, longer monitoring periods will be crucial. Managers will have to consistently train data collectors and maintain formats and collection points to assure consistency over time. However, once consistent monitoring protocols are established, managers can implement policies designed to enhance the river system's resilience and will be able to evaluate the impacts of those policies over time.

Conclusions and key lessons

Using an example of monitoring the three forks of the Flathead River System in Montana, this chapter shares the diverse tools available to monitor use on and off the river. The strengths and limitations of monitoring and analysis strategies are discussed in addition to examples of how to synthesize and compare data. In this final section, we offer recommendations for research topics to stimulate future growth in monitoring and managing a river system for sustainable tourism in a complex social-ecological river system.

- *Understanding the key factors to successful collaboration for a transboundary resource*: The transboundary nature of rivers involves collaboration across agencies and partners that is representative of stakeholders from the local, state and national scales in the planning process. Integration and sharing of data across partners is necessary, but can be difficult as different agencies and partners collect and manage data in diverse ways. Additionally, inconsistent budgets can pose difficulty for long-term planning and monitoring. The Flathead River planning process has been successful in leveraging support and resources across partners and is likely going to be necessary in long-term management of the river system.
- *Diversifying perceptions of the river user experience*: Perceptions of experience can differ from on-shore and on-river. Standards and indicators are

required for both and need to be updated through surveys as experiences and resources change. There is no single "best" monitoring method; rather, careful thought and analysis are needed. Different data collection methods are complementary in gaining a comprehensive understanding of recreational river use. There are strengths and challenges with both monitoring approaches relevant to using one approach alone. The USFS data is limited in scope, but critical to understanding the on-river users' experience. The on-shore collection method shows the on-shore perspective, but is limited to stationary locations along the river.

- *Utilizing diverse methods to maximize accuracy and efficiency*: These two methods also vary in precision of data collected; the USFS monitoring provides precise information regarding launch site wait time on specific days, as well as how many shore users are along a stretch of river on that day. The on-shore data provides precise information regarding the number of on-river users who pass by a certain site, as well as providing information at a site for every day of the season. The efficiency of these methods varies whereas the USFS on-river monitoring requires river rangers and volunteers to conduct full day river floats which is time consuming and causes the number of stretches monitored to be limited to one per day. However, data entry is more streamlined in this method as paper forms only need to be entered into the computer following the float. The on-shore monitoring is less time consuming in the data collection process because cameras only need to be serviced every two to three weeks. In particular, efficiency is lost in on-shore monitoring in the data processing where watching of river footage is extremely time consuming.
- *Making the most of data and integration in diverse and complementary planning processes*: The collection, management and analysis of data are crucial. However, just collecting these different data types is not enough. It is important to have a strategy for how the information will be synthesized into the management and planning of the system. In the case of the Flathead River System, this means incorporating the different types of data into the Comprehensive River Management Plan process. The information provided by the data not only needs to be understandable to managers for making decisions. It also needs to be synthesized by researchers and managers in a way that is digestible for the public to understand why this data is being used to inform management decisions. This could mean presenting data trends and findings to the public through community meetings.
- *Long-term monitoring of trends*: The collection and analysis of data over time is an important component of the monitoring process. Data fidelity is crucial to documenting and understanding the changes in use and the resource over time. Without the evidence the numbers bring, it will be even more difficult to understand the complex system involved in river use and tourism.
- *Integration of diverse forms of data and new innovative data and analysis techniques*: It is important to consider and integrate many different types of data. Unfortunately, there is no single method by which to integrate different types and sources of data and each format must be considered on a case by case

basis. Rather, we should think of integration as assembling evidence to see the connections, feedbacks and complexities in the system.

These recommendations illuminate the unique managerial characteristics of a river system and in particular a Wild and Scenic River. The Flathead River System's growth in use has been influenced by an influx of residents to the region and increasing tourism growth to Glacier National Park and northwestern Montana. Additionally, ecological changes in climate that have led to extreme temperatures, dry conditions and intense wildfires in the summer which have influenced use and visitation to the river and surrounding region.

To manage a river that encompasses one of the most visited national parks as well as a wilderness area, monitoring is critical to respond to changes and assess whether standards are being maintained for resources and experiences. A resilient management system can monitor and anticipate changes or disturbances to the resource or experience. Managers can then adapt and respond efficiently and effectively to ensure that the resource or experience can maintain their character based on the indicators and standards in place.

This chapter illustrates the complexities and challenges of monitoring a river system and also offers strategies for utilizing different methods, interpreting and synthesizing data, and developing a highly effective management strategy. Norman Maclean (1976), author of an influential novel, *A River Runs Through It*, writes, "Eventually, all things merge into one, and a river runs through it." The Flathead River, similar to other rivers, is a central part of social-ecological systems and without complex monitoring strategies, the resiliency and adaptive management of the system will be compromised.[2]

References

Allen, C. R., Fontaine, J. J., Pope, K. L. & Garmestani, A. S. (2011). "Adaptive management for a turbulent future." *Journal of Environmental Management*, **92**(5): 1339–45.

American Rivers (2017). "What is a wild and scenic river?" Retrieved October 1, 2018 from https://www.americanrivers.org/threats-solutions/protecting-rivers/.

Ames, R. T. & Schindler, E. (2009). "Video monitoring of ocean recreational fishing effort." Oregon Department of Fish and Wildlife, Information Reports No. 2009-01.

Auerbach, D. A., Deisenroth, D. B., McShane, R. R., McCluney, K. E. & Poff, N. L. (2014). "Beyond the concrete: accounting for ecosystem services from free-flowing rivers." *Ecosystem Services*, **10**: 1–5.

BLM Spokane District Office (2010). "Outstanding remarkable values (ORV) criteria and regions of comparison." Retrieved September 25, 2018 from https://www.blm.gov/or/districts/spokane/plans/ewsjrmp/files/WSR_handout_02.pdf.

Chaffin, B. C., Gosnell, H. & Cosens, B. A. (2014). "A decade of adaptive governance scholarship: synthesis and future directions." *Ecology and Society*, **19**(3). http://dx.doi.org/10.5751/ES-06824-190356.

Chamberlain, L. (2016). "GeoMarketing 101: What is geofencing?" Retrieved November 20, 2018 from https://geomarketing.com/geomarketing-101-what-is-geofencing.

2 The authors gratefully acknowledge the support and input of the USFS, the NPS, and the Glacier Park Conservancy.

Clark, R. N. & Stankey, G. H. (1979). "The recreation opportunity spectrum: a framework for planning, management, and research." General Technical Report PNW-98. Retrieved November 20, 2018 from https://www.fs.fed.us/cdt/carrying_capacity/gtr098.pdf.

Edwards, J. & Schindler, E. (2017). "A video monitoring system to evaluate ocean recreational fishing effort in Astoria, Oregon." Oregon Department of Fish and Wildlife. Retrieved from https://www.recfin.org/wp-content/uploads/2017/06/Astoria-VBC-Project-Final.pdf.

Folke, C., Carpenter, S. R., Walker, B., Scheffer, M., Chapin, T. & Rockström, J. (2010). "Resilience thinking: integrating resilience, adaptability and transformability." *Ecology and Society*, **15**(4). http://www.ecologyandsociety.org/vol15/iss4/art20/.

Great Falls Tribune (2014). "Ultralight packrafts make way for new genre of outdoor adventuring." Retrieved November 1, 2018 from http://www.spokesman.com/stories/2014/aug/24/ultralight-packrafts-make-way-for-new-genre-of/.

Greenberg, S. & Godin, T. (2015). "A tool supporting the extraction of angling effort data from remote camera images." *Fisheries*, **40**(6): 276–87.

Herrera Environmental Consultants (2014). "Synthesis of 2013 River Recreation Studies: King County River Recreation Study" (Rep.). (October 31). Retrieved November 4, 2018 from https://your.kingcounty.gov/dnrp/library/2014/kcr2629.pdf.

Hirabayashi, Y., Mahendran, R., Koirala, S., Konoshima, L., Yamazaki, D., Watanabe, S., . . . & Kanae, S. (2013). "Global flood risk under climate change." *Nature Climate Change*, **3**(9): 816.

IVUMC [Interagency Visitor Use Management Council] (2016). "Visitor Use Management Framework: A guide to providing sustainable outdoor recreation." Retrieved October 2, 2018 from https://visitorusemanagement.nps.gov/Content/documents/lowres_VUM%20Framework_Edition%201_IVUMC.pdf.

Long, J. S. (2009). *The Workflow of Data Analysis Using Stata*. College Station, TX: The Stata Press.

Maclean, N. (1976). *A River Runs Through It*. Chicago: University of Chicago Press.

National Park Service (2006). "Colorado River Management Plan Record of Decision." Retrieved October 17, 2018 from https://www.nps.gov/grca/learn/management/upload/CRMP_ROD_2006.pdf.

National Park Service (2018). *Glacier National Park YTD Report*. Accessed on October 12, 2018 from https://irma.nps.gov/Stats/SSRSReports/Park%20Specific%20Reports/Park%20YTD%20Version%201?Park=GLAC.

Postel, S. (2014). *The Last Oasis: Facing Water Scarcity*. Abingdon: Routledge.

Powers, S. P. & Anson, K. (2016). "Estimating recreational effort in the Gulf of Mexico Red Snapper Fishery using boat ramp cameras: reduction in federal season length does not proportionally reduce catch." *North American Journal of Fisheries Management*, **36**(5): 1156–66.

Stankey, G. H., Cole, D. N., Lucas, R. C., Peterson, M. E. & Frissell, S. S. (1985). "The limits of acceptable change (LAC) system for wilderness planning." General Technical Report INT-176. Retrieved November 20, 2018 from https://www.fs.fed.us/cdt/carrying_capacity/lac_system_for_wilderness_planning_1985_GTR_INT_176.pdf.

Steele, B., Chandler, J. & Reddy, S. (2016), *Algorithms for Data Science*. Dordrecht: Springer.

Taber, B. N. (2012). "Recreation in the Colorado River Basin: is America's playground under threat?" Retrieved October 15, 2018 from https://www.coloradocollege.edu/dotAsset/c1d0b548-4350-4be7-b0a5-8de6692b973b.pdf.

Tufte, E. (1983). *The Visual Display of Quantitative Information*. Cheshire, CT: Graphics Press.

U.S. Census Bureau (2017). "County population by characteristics: 2010–2017." Retrieved October 20, 2018 from https://www.census.gov/data/datasets/2017/demo/popest/counties-detail.html, 2017.

U.S. Forest Service (1980). *Flathead Wild and Scenic River Management Plan*. United States Department of Agriculture.

U.S. Forest Service (1986). *Flathead Wild and Scenic River Recreation Management Direction*. United States Department of Agriculture.

U.S. Forest Service (2018). *2017 North Fork and Middle Fork Flathead Wild and Scenic River System Annual Monitoring Report*. United States Department of Agriculture.

USFS & NPS (2013). "Flathead National Forest & Glacier National Park. Outstandingly remarkable values assessment of the Flathead River System." Retrieved from https://www.fs.usda.gov/Internet/FSE_DOCUMENTS/fseprd570451.pdf.

Wenger, S. J., Isaak, D. J., Dunham, J. B., Fausch, K. D., Luce, C. H., Neville, H. M., Rieman, B. E., Young, M. K., Nagel, D. E., Horan, D. L. & Chandler, G. L. (2011). "Role of climate and invasive species in structuring trout distributions in the interior Columbia River Basin, USA." *Canadian Journal of Fisheries and Aquatic Sciences*, **68**(6): 988–1008.

13 Communities and sustainable tourism development: community impacts and local benefit creation in tourism

Jarkko Saarinen

Introduction

In development discourses, the tourism industry is often seen as a beneficial tool for local development, serving as a response to negative economic changes taking place in rural areas, especially, but also in urban environments. Recently, the World Bank Group (2017) launched their list of 20 reasons why tourism works for development. These reasons are positioned within a sustainable tourism framework and the United Nations Sustainable Development Goals (UN SDGs), under the five following elements: Sustainable economic growth; social inclusiveness, employment, and poverty reduction; resource efficiency, environmental protection, and climate; cultural values, diversity, and heritage; and mutual understanding, peace, and security. The emphasis of these elements and the specific reasons for supporting the tourism industry in global development thinking are largely based on positioning the industry's role in a place-based context as a local solution for global challenges.

Thus, the emphasis of the beneficial role of the tourism industry for development is typically placed on a local scale, in community contexts: The growth of the global industry and its capacity to assist communities, their development and well-being in destination regions is expected to trickle down. In this way, tourism is framed as a sustainable development path for local communities, and the very idea of sustainability has been incorporated into the tourism industry's policies and development and governance thinking (Bramwell 2011; Saarinen, Rogerson & Hall 2017). This has led Hall (2011: 650) to state that the idea of sustainability is "one of the great success stories of tourism research," making tourism and related studies highly policy-relevant issues in development discussions. Recently, the United Nations Research Institute for Social Development report *Revisiting Sustainable Development* (UNRISD 2015) has similarly highlighted the global policy success of the overall idea of sustainability: "It seems that virtually all development actors and organizations, and the public at large, have bought into the narrative of sustainable development" (Utting 2015: 1). However, as further noted by Utting (2015: 1) in the UNRISD report, the core elements and ideas of sustainability "often got lost in translation" when we are "doing development," that is, we too often emphasize the short-term economic dimensions over social and environmental ones.

Unfortunately, this general characterization applies to tourism and its developmental role in community contexts (Sharpley 2000; Liu 2003; Saarinen 2014).

This chapter focuses on tourism and community relations in the context of sustainable development. In general, host–guest relations and tourism impact for communities are issues that have been widely studied and discussed in tourism research (see Murphy 1985; Smith 1989). Here the aim is to overview the role of tourism in community development and discuss the complexities of local benefit creation in tourism development. First, the idea of sustainable tourism is outlined from a community perspective, followed by a discussion on the tourism impact on communities. After that, some of the key elements influencing benefit creation in tourism and community relations are discussed and summarized based on literature. Finally, future research needs and development questions are highlighted in tourism and community relations.

Contextualizing sustainable tourism in community development

Sustainable tourism has been defined by and approached from various perspectives (see Sharpley 2000; Saarinen 2014; McCool & Bosak 2016). Perhaps the most commonly used definitions originate from policy documents created by the World Tourism Organization, which considers itself as being "responsible for the promotion of responsible, sustainable and universally accessible tourism" (World Tourism Organization 2016). Their early conceptual formulation stated that sustainable tourism aims to meet "the needs of current tourists and host regions while protecting and enhancing opportunities for the future" (World Tourism Organization 1993: 7). Later the UNWTO and the United Nations Environment Programme (UNEP) jointly defined sustainable tourism as "tourism that takes full account of its current and future economic, social and environmental impacts, addressing the needs of visitors, the industry, the environment and host communities" (UNEP 2005: 12). Both policy-oriented definitions highlight the role of hosts, that is, destination communities and their living environment. However, it is notable that compared to the original definition from the early 1990s, the specific emphasis on meeting the "needs of the industry" has been added to the latter definition. Currently this industry orientation is a characterizing approach in organizing sustainability (see United Nations Secretary-General's High-Level Panel on Global Sustainability 2012; World Tourism Organization 2016; World Bank Group 2017) and it characterizes many academic conceptualizations as well. However, industry-driven sustainability can be highly problematic if the needs of the industry, hosts and/or environmental protection needs conflict with each other (Hunter 1995; Sharpley 2000; Liu 2003); a situation that is not highly unlikely in tourism–community–environment relations.

This kind of economic emphasis controverts a common public opinion that sustainability is specifically an environmental issue dealing with impacts and changes in ecosystems, habitats and species. However, the original definition of sustainable development did not highlight the environment in the first place but the needs of the current and future generations (i.e., people). This focus on people was also mentioned in the World Conservation Strategy in 1980 (IUCN 1980: 7) that preceded Brundtland's influential report "Our Common Future" in 1987 (WCED 1987).

In this complex set of relations between environmental, social and economic dimensions, the core question is: Whose needs should guide the setting of the limits to growth in tourism? In an ideal world the limits would probably be adaptively defined based on the most threatened and vulnerable elements in the destination regions. That way the local development would have a better opportunity to meet the needs of present generations without endangering the ability of future generations to meet their own needs. In reality, however, current market-driven neoliberal approaches have often positioned the industry and its economic growth needs in a hegemonic role when thinking about and defining the limits to growth and the desired goals for (tourism) development in destination regions. Thus, the economic pillar is often considered taller than the environmental and social ones. In practice, this may place an explicit or implicit emphasis on sustaining the industry and its resource needs in (sustainable) tourism development (Saarinen 2014). Obviously, this can be challenging for the social and ecological sustainability of a host region. In addition, it can also be economically challenging for a region, as tourism may not always be the most sustainable way to use local natural and/or cultural resources (Butler 1999).

In contrast to tourism-centered sustainability in planning and development, the industry should be seen as a potential tool for sustainable development (Burns 1999). That way, sustainable tourism would represent "a subset of sustainable development" and a "system that encourages qualitative development, with a focus on quality of life and wellbeing measures" (Hall, Gössling & Scott 2015: x). Obviously, a key issue is how we should understand the idea of development and its relationship with growth in local and community settings (Daly 1996). Although there is no consensus about what development or growth precisely mean (Saarinen, Rogerson & Hall 2017), their difference is well acknowledged in tourism studies (see Telfer & Sharpley 2008; Scheyvens 2011). This difference can be demonstrated by utilizing the United States Local Government Commission's definition, which indicates that "growth means to get bigger, development means to get better – an increase in quality and diversity" (cited in Pike, Rodríguez-Pose & Tomaney 2007: 1253). Thus, growth is a quantitative indicator, while the idea of development is more focused on qualitative dimensions in social and economic processes (see Hall, Gössling & Scott 2015). Therefore, development refers to quality of life and well-being in community settings. Current development and tourism literature also uses the term inclusive growth, which represents an alternative view that emphasizes the developmental aspects and potentials of tourism growth for destination communities and their quality of life and well-being (see Scheyvens 2011; Butler & Rogerson 2016). Inclusive growth in tourism has the capacity to complement other local development strategies, which are targeted at fostering economic diversification and transformation in rural and peripheral settings (UNCTAD 2017; Rogerson

& Saarinen 2018). Whether we use the term development or inclusive growth, the key issue making tourism more sustainable for communities is that there are not only economic but also social and ecological considerations involved in development, that communities have a say and that there are supportive governance structures for defining and setting limits to growth and change in tourism (see McCool, Moisey & Nickerson 2001; Mosedale 2014). The question is how to govern the impacts of tourism development, and who does it.

Communities and tourism impacts

Tourism comes with impacts. In the context of sustainable development and its basic elements, the impacts of tourism are broadly categorized into environmental, social and economic. On a general level, these different impacts of tourism can be positive or negative for communities or specific segments of local populations (see Mathieson & Wall 1982). This means that the economic effects, which are usually regarded as positive per se, can also be negative for communities. Similarly, the environmental impacts of tourism can be positive ones. For example, growing tourism activities may support nature conservation initiatives, which is obviously a good trend for the environment and people working in nature-based or ecotourism sectors. However, these initiatives can cause negative economic outcomes for local resource-extractive industries and supporting businesses and, as a result, those people employed in the sectors. In addition, tourism impacts can be direct or indirect in space and time, that is, some impacts can materialize in a different location from that which tourists actually use, and/or impacts evolve and appear much later in a socio-ecological system. Destination communities are also linked to wider tourism and socio-economic, environmental and political systems that influence and cause impacts on destinations. These different types and scales of tourism impacts are often interrelated, the complexity of which makes it very challenging (but also interesting!) to study them and analyze their possible linkages and causal relationships in destination communities and beyond.

From a community perspective, all tourism impacts are relevant to people. Thus, environmental and economic changes caused by evolving tourism also have social meanings and functions for local people. Here, however, for the sake of clarity we focus mainly on the social impacts of tourism. According to Hall (1994) the social impact of tourism refers to the ways in which the industry and tourists create changes in communities; to individual and/or collective behavior, lifestyles, values, and quality of life. There are many elements influencing how local residents perceive these changes taking place in their everyday lives, such as the nature of tourism development and tourist activities and behavior, community characteristics, local economic dependency on tourism, and the scale of tourism (Butler 1980; Murphy 1985; Nunkoo & Ramkissoo 2011; Nunkoo & Gursoy 2012). One of the best-known frameworks to demonstrate different community attitudes toward tourism is based on Doxey's (1975) Irritex (irritation index) model. The basic idea of the model is that, as the industry grows, local attitudes change from "euphoria" to "apathy," followed by "annoyance" and "antagonism." These different phases of attitudes reflect community responses to the changing impacts of tourism, specifically toward negative externalities resulting from tourism-based growth and activities. A similar approach is Butler's (1980) Tourism Area Life Cycle (TALC) model, which focuses more on the changing impacts and nature of tourism activities, but the model can also involve local responses and attitudes toward the transforming characteristics of a tourist destination (see Butler 2006).

A common approach in various academic models reflecting the host–guest relations is the doom and gloom view. Based on this approach, an initially positively perceived tourism will eventually be seen as ruining the place and turning local attitudes negative toward the industry (see Murphy 1985; Gursoy, Chi & Dyer 2009; Kim, Uysal & Sirgy 2013). This deterministic line of thinking may not always fully capture the community dynamics in tourism (see Ma & Hassink 2014), but it is a common view of tourism–community relations in research and also characterizes popular views and discourses on tourism development, especially in relation to international (mass) tourism. Recent examples include debates and protests focusing on the process of overtourism in various global (tourist) cities, such as Barcelona, Berlin, London and Venice (see Paasi et al. 2019). In these discussions the overtourism phenomenon seems to reflect a crisis of contemporary tourism and the ways the host–guest relations are transforming, due to new modes of tourism production and their connections to housing markets and gentrification processes in various metropolitan locations. Thus, the issue is not simply about the scale of tourist mobility or overcrowding. Obviously, there are too many visitors in particular destinations at the same time—an "overdose of tourists"—but the overtourism discourse is also about the ways in which the industry works by utilizing the sharing economy, digitalization and related platforms, for example, and how that process changes urban spaces and the nature of tourism impacts in neighborhoods where local people live their everyday lives (see Nofre et al. 2018). These individualistic "new tourists 2.0" of the 2010s (compared to the "new tourists" of the 1990s as postmodern consumers, see Urry 1990; Poon 1993) utilize Airbnb, Couchsurfing and other such platforms for accommodation and related services in everyday residential areas. By doing so they are "living like locals" and are thus increasingly entering and occupying spaces that are usually seen as non-touristic; the domain of the everyday lives of local people. All this challenges the borders between hosts and guests (see Paasi et al. 2019), the distinction of which is symbolically captured in the application of Goffman's (1959) sociological conceptualization of front- and backstages in tourism studies (MacCannell 1976); the frontstage is for touristic displays, while the backstage involves authentic, local ways of living. By entering the backstage in large numbers, tourists are rapidly changing local urban lifestyles and environments, thus causing annoyance and antagonism in Doxey's (1975) terms.

However, while the industry may have the capacity to destroy its destinations, the development of tourism does not only involve costs but also benefits. Table 13.1 summarizes some of the key social impacts of tourism for local communities. Although the table is divided categorically into positive and negative impacts, it is

Table 13.1 Social impacts of tourism: the main benefits and costs of tourism for communities

Benefits: positive impacts	Costs: negative impacts
Employment and income	Resource and land use conflicts
Diversification of local economy	Increased inequalities
Well-being/quality of life	Higher living costs and inflation
Poverty alleviation	Crime and other social problems
Better infrastructure and services	Acculturation
Revitalization of culture	Commodification of traditions
Cultural learning	Demonstration effect, change of lifestyle
(Resource) conservation	Seasonal and low-paid employment

important to note that some of the listed impacts can represent both benefits and costs for a community and/or its different sub-groups and individuals. Indeed, communities are not socially homogeneous entities but, in general, the members of a community have an understanding of shared elements of identity and norms, and they interact with each other based on certain kinds of norms and behavioral models in reciprocal ways. In addition, communities are usually attached to a certain socio-spatial unit, such as a village, town or suburb. Living in the same space, however, does not automatically make people a community, but the places people live in often form a basis for shared social identities, interaction and norms guiding the use and sharing of local resources and attitudes toward tourism, for example.

As indicated in numerous policy documents and empirical studies on the impacts of tourism, there is a multitude of evident and potential benefits for communities. Compared to many other economic sectors, tourism is often a relatively effective and inexpensive tool for creating employment (Sinclair 1998; Scheyvens 2002). In addition to direct and indirect employment opportunities, tourism provides valuable income, and the industry can be used for local and regional socio-economic diversification purposes (Telfer & Sharpley 2008; Saarinen & Lenao 2014). The industry can also be used for the restoration of cultural and natural heritage resources and the improvement of local services and infrastructure (Smith 1989; Hall 2008). Furthermore, the evolution of tourism can diversify the demographic basis of the destination communities (Murphy 1985; Butler 2006).

However, tourism development comes with its own set of risks (World Bank 2012) and, thus, involves potential negative impacts, especially if it is not well planned and managed (Hall 2008). Indeed, tourism reportedly brings about various social problems, including crime, prostitution, and displacement (see Mathieson & Wall 1982; Mbaiwa 2005), and while tourism development may create more employment opportunities, the jobs are often seasonal and low paid. This, together with increasing living costs and inflation due to non-local consumption, will lead to increased inequalities in communities and host–guest relations (Hunt et al. 2015). In addition, tourism can cause negative changes to well-being and lifestyles, and tourism producers can promote and commodify local cultures and traditions in unethical ways (Smith 1989).

Thus, there are both social benefits and costs involved with tourism development. In the context of sustainable development in tourism, the underlying idea is to work with the costs and benefits by minimizing the former and maximizing the latter. This balanced approach is widely discussed in tourism studies (see Bramwell 2011; Hall 2011). Next, the key elements influencing and guiding community benefits creation in tourism will be overviewed and discussed.

Communities and benefit creation in tourism

There has been considerable scholarship on tourism impact and benefit creation in tourism (Hunt et al. 2015). Studies on the impact of tourism have identified a broad range of effects for communities and have indicated elements and issues that can play a major role in how the impacts occur and how communities could become agents and subjects of change that involves tourism development (Murphy 1985; Nunkoo & Gursoy 2012). From a sustainable development perspective, the question is how communities could optimize the benefits and effectively manage the potential costs of tourism. Based on tourism research literature, the following elements can be highlighted: Participation; control; tourism awareness and knowledge; community characteristics; and governance structures. These elements are analytic by nature, that is, in practice they are often intertwined in complex ways in tourism planning and development processes.

Participation: The need to be involved with tourism

Community participation is seen as the key element for sustainable tourism planning and development (Scheyvens 1999; Tosun 1999; Valentin & Spangenberg 2000; Li 2006). Participation represents the core issue in benefit creation but, as noted by several authors, it is often vaguely used in tourism studies (Timothy 1999; Scheyvens 2002; Tosun 2006). In general, it "refers to a form of voluntary action in which individuals confront opportunities and responsibilities of citizenship" (Tosun 2000: 615). In tourism, the term can be "understood as the involvement of individuals within a tourism-oriented community in the decision-making and implementation process with regard to major manifestations of political and socio-economic activities" (Pearce, Moscardo & Ross 1996: 181).

Put simply, in order to benefit from tourism, community members need to be involved in the industry. However, this involvement of individuals can take various forms and levels. Tosun (2006) has classified the involvement into coercive, induced and spontaneous community participation. Briefly, coercive participation is a formal top-down approach in which participation takes place in the implementation phase of tourism planning and development, but it does not necessarily lead to real benefit sharing. Instead, it may primarily benefit and serve the needs of the industry and other decision-makers mainly operating outside of the involved community. Induced participation takes a step in a consultative direction, in which community members are heard, especially in the implementation phase of the development process. Still, although host communities are symbolically heard, their role is often characterized by being passive recipients (that is, objects) of tourism development and tourism-based benefits. Thus, community members "are heard but not listened to," and they do not set the priorities and goals for (tourism) development. Contrast these approaches with spontaneous participation, which is based on an active role of community members with a bottom-up approach and direct participation in actual decision-making, including the priority setting processes in the industry and/or resource and land uses. Thus, a community is an active agent and subject in planning and development.

Control: The issue of power

While community participation is key in receiving and creating benefits from tourism, it is clear that a mutually beneficial participation process needs other qualities than simple involvement alone. Based on the literature, participating individuals and communities need to have mechanisms to influence the decision-making processes in tourism planning and development (Tosun 2006; Hunt & Stronza 2014). Thus, communities should have a certain level of control over tourism activities taking place and using resources in their living environment. The question is about power and power relations in the tourism–community nexus, which are often problematic issues between private sector industry, local institutions and people (Church & Coles 2007; Hall 2008).

The idea of sharing and devolving power is built into the community participation theories in development studies. For example, Arnstein (1969: 216) has defined participation as "the redistribution of power that enables the have-not citizens" to have a say and share "the benefits" of development. Also, many Community-Based Natural Resource Management (CBNRM) models (see Blaikie 2006), in which tourism planning and development projects are often linked in rural areas (van der Duim et al. 2012), are based on the devolution of power from central government and industry to communities. In tourism planning and development contexts, the sharing of power and, thus, supporting benefit creation from tourism for communities can be organized by creating joint ventures and institutional partnerships, for example, between communities and the private sector (Kavita & Saarinen 2016). In addition, local control mechanisms can be based on participatory planning, collaborative management and communication systems, land and other resource leasing systems or a combination of them (Jamal & Stronza 2009). In an optimal situation, these kinds of mechanisms generate self-determination that enables communities to represent their interests with their own authority in planning and development processes. This community self-determination is the grounding for local empowerment, which Scheyvens (1999) has divided into economic, psychological, social and political empowerment (see also Scheyvens 2002). The key issue in community empowerment is the idea of agency: Instead of objects, local community members are positioned as subjects in tourism planning and development processes. The community's position often defines the limits and nature of their participation in tourism.

Tourism awareness and knowledge

In order to participate in beneficial but also meaningful ways in tourism, communities need to feel empowered (Scheyvens 2002). In addition, they also need to have an understanding of how tourism works and how the tourism system is organized in a global–local nexus. This calls for awareness and knowledge of tourism, its impacts and the community's role in the industry's circle of production. As a term, tourism awareness refers to the level of local knowledge about tourists, the tourism system, and its impacts—issues that are often neglected in development and community participation discussions (Saarinen 2010). However, knowledge differences between industry actors and local people may lead to the cultural limits of participation (Tosun 2000), which can deepen inequalities between hosts and guests in tourism.

Novelli and Gebhardt (2007: 449) have further pointed out that local communities need to achieve "similar levels of understanding and knowledge" with other stakeholders in order to fully participate in tourism development processes (see also Reed 1997). Thus, participation is not only about achieving a more equitable distribution of material benefits but also sharing knowledge and learning (McCool 2009). In this respect, a lack of education and training may create constraints for local people to participate and therefore limit the benefits of tourism. Related to this, Reid (2003) has strongly emphasized the need to raise community awareness in terms of collaborative tourism development. Many CBNRM projects also highlight the need for local training and education in tourism and conservation projects (Scheyvens 2002). However, this training does not need to be limited only to locals. The industry and development agencies should learn how communities operate with their resources and what the community preferences and dynamics are in destination environments.

Community characteristics

As indicated earlier, communities are not homogeneous units. This notion applies to the internal relations and sub-groups of communities, but also to wider characteristics of communities. This means that communities may cope very differently with evolving tourism activities, their impacts, and management. In the context of indigenous tourism, for example, it has been indicated that cultural groups differ in respect of their acculturation process (Butler & Hinch 2007): Some communities may have cultural traditions and norms that empower them to resist or cope with changes and influences originating from tourism, while other communities can more easily adopt external values, norms and practices from other cultures. In tourism, these other cultures usually refer to visiting tourists. Communities may also have different ways of using local resources, which can influence the host–guest relations in practice. If the use of resources in tourism supports traditional local livelihoods, for example, the adaptation of touristic views and acculturation can take place more quickly in practice compared to conflicting situations in resource utilization.

One element that may explain different capacities to resist cultural changes (or learning) is the idea of social capital (Jones 2005). Social capital has numerous different definitions and meanings (see Bourdieu 1986; Putnam 1995) but, in general, it involves social relations that are characterized by the norms of reciprocity and trust between community members (Putnam 2001: 19). Tighter relations, trust and norms lead to stronger capacity to maintain a community's own way of thinking and governing local socio-economic processes. In this respect, the role of trust has recently been highlighted in research literature (see Nunkoo & Ramkissoo 2012; Nunkoo & Smith 2013).

Thus, in addition to the nature, phase and pace of tourism development, community characteristics and social capital influence local attitudes toward tourism and its impacts.

Governance: The regulative landscape and institutional arrangements in the policy arena

The idea of social capital refers to the organization and dynamics of community governance. However, communities or tourism planning and development processes do not happen in a societal vacuum. Instead, they take place and are integral parts of larger socio-ecological, political and economic systems. How these relationships between larger and local systems are organized is a matter of governance and related institutional and societal arrangements, which can favor communities' roles or create disadvantages for them by preferring the industry or some elite groups in development, for example (see Tosun 1999). Governance involves public and private sector actors, including formal institutional structures, businesses, non-governmental organizations and other groups that form society in the increasingly transnational context of tourism planning and development (Stoker 1998). In principle, preferred and utilized modes of governance indicate our understanding of who should steer societal development and how. On the one hand, governance can be based on an active role of formal institutions and government. On the other hand, governance refers increasingly to neoliberal market-oriented reform policies that aim to reduce the role and influence of the state on the economy and development, especially through privatization (Jessop 2004; Harvey 2005).

Thus, the issue of governance in tourism, with its connections to communities, government and non-governmental agencies, is a highly challenging task and field of analysis. For Rhodes (1996: 653), one key aspect of governance is about allocating societal resources with a set system of control and coordination. This way, the idea of governance involves previously outlined power issues and trust in tourism and community relations, but it is also related to sustainability management in tourism planning and development. As noted by Bramwell (2011: 461): "Destinations wanting to promote sustainable tourism are more likely to be successful when there is effective governance." In relation to the aims and ideology of sustainable development, this referred effective governance is increasingly understood to be a call for a regulative framework that would be based on relatively strong governmental and/or inter-governmental policies and regulations (see Jessop 2010). Although it is not particularly fashionable in the current context of neoliberal politics, this call for stronger governance is evident in respect of climate change policies, resilience and poverty alleviation in tourism and community relations, for example (see McKercher et al. 2010; Jamal 2013).

Research needs in tourism, community benefits and sustainability

Benefit creation in tourism is a complex set of elements and dynamics taking place in a local–global nexus, creating a challenging system for sustainable tourism governance (Bramwell & Lane 2011; Hall 2011). In that system, in general, the benefit creation needs to outstrip the costs caused by tourism. Indeed, as indicated by Sinclair (1988) the positive (economic) aspects of tourism should be placed in an equation consisting of both the advantages and disadvantages of tourism-related development (see Rogerson & Saarinen 2018). Even the World Bank (2012: 7), which is usually seen as a lobbying organization for socio-economic growth opportunities linked to the tourism industry, has stated that tourism "comes with its own set of risks and challenges."

However, the issue is not only the "balance" between the costs and benefits of tourism development for communities, but also the nature of those benefits: Not all community benefits are necessarily in line with the premises and needs of sustainable development in tourism (Butler 1999). Too often, short-term economic dimensions are emphasized over environmental and social ones (Hunter 1995; Holden 2018). In this respect the key questions to ask are: What does sustainability mean for community development, and what is it that tourism should sustain? (McCool & Bosak 2016). Obviously, these are community-specific questions to which the answers depend on the communities and their needs within a socially, economically and environmentally sustainable tourism development process. Recent literature has highlighted the idea of community resilience as a target for sustainable tourism development. The concept originates from ecological studies (Holling 1973) and it has been applied to human systems by defining resilience as community's "capacity to rebound to a desired state following both anticipated and unanticipated disruptions" (Lew 2014: 14). As an approach, resilience calls for system thinking and community adaptability within a larger and complex socio-ecological context.

Sustainability and resilience are associated terms in tourism and community planning and development. Some scholars have further speculated whether resilience could replace the idea of sustainability in development discourses and management (see Lew 2014; Espiner, Orchiston & Higham 2017). While sustainability as an ideologically grounded goal for humanity has its issues as an academic concept with challenges in terms of practical implementation, resilience itself may not be the panacea or holy grail for sustainable development. Being resilient in currently changing environments does not automatically lead communities onto a sustainable development path, as their capacity to resist and cope with change can be based on a locked-in situation with unsustainable ways of living and using resources in the long term (Martin & Sunley 2014). Therefore, community resilience needs to be integrated—as a tool and/or a goal—into the overall process of sustainable development in tourism within wider socio-ecological system thinking.

Obviously, there are many changes that will take place in tourism and community planning and development contexts in the future, which are highly challenging to predict and consider proactively. In addition to the changes in the natural environment and the increasing impacts of the Anthropocene (Latour 2015; Moore 2015), for example, there will be technological changes that can lead to significant transformations in host–guest relations and that cause major changes to how communities are impacted by tourism and how they can participate in and have control over tourism development in the future. The issue of overtourism penetrating the backstages of host communities is partly based on such technological advancements in the form of the sharing economy. Technological changes will deepen and transform customer, provider and business-to-business relations in unimagined ways in the future. For example, the Fourth Industrial Revolution (IR4.0) in the tourism and hospitality sector is connected with some of the key trends in future tourism: The need for sustainability in tourism growth, changing mobilities, digitalization, and enabling technologies. The IR4.0 approach can provide improved efficiency for tourism and hospitality operations, possibilities for new kinds of products, and greater personalization of tourist experiences, enabling connectivity between people and with the environment. This "fusion of technologies blurs the lines between the physical, digital, and biological spheres" in tourism experience and management (see Schwab 2015), which will also create changes and potential challenges for local participation and well-being in tourism development. According to Othman, Ramli and Ahmad (2017), there is a need for transformative support in the traditional small- and medium-sized tourism and hospitality sectors to help them adopt a new "culture," and change and risk management thinking in the use of new technologies and related business models. In addition, sustainability management in IR4.0 development environments will require training, new kinds of monitoring systems, and well-balanced governance models between businesses, communities and the public sector. All these emerging integrative and novel uses of technology call for a new kind of interdisciplinary research in sustainable tourism governance in community settings.

Summary and a way forward

It is clear that tourism can have a major impact on destination communities. Questions about how community benefits are created and how they can be optimized in tourism planning and development are important to consider. The industry has the capacity to create various kinds of benefits for communities if its planning and development processes are well governed. However, characteristically the transnational tourism system and its relationship with the socio-ecological systems in destination areas and beyond are highly complex. Therefore, tourism–community relations are challenging to govern and deconstruct in tourism management and research. Still, there is a real need to actively consider and understand how a constantly evolving tourism industry operates in destination communities and how it could be steered onto a sustainable development path. In this respect, one critical element is the growth-orientation of the global tourism industry.

This growth-orientation of the industry creates the potential for using tourism for development in various community settings. This means that many development agencies consider tourism as an effective tool for poverty alleviation (see World Bank Group 2017). However, the current growth ideology of tourism is fueled by neoliberal thinking, which may not create development but simply increase inequalities in different localities, as noted by Ostry, Loungani and Furceri (2016). The growth of tourism can also cause various negative externalities for communities and their everyday living and working environments. Indeed, while highlighting the prospects based on the increasing numbers of (international) tourist arrivals, we tend to forget that the growth of tourism does not necessarily equal sustainability of the industry or the idea of development equaling well-being and quality of life. Thus, due to the growth- and market-driven nature of the tourism industry, the need for sustainability has become highly crucial and probably the greatest challenge of tourism, leading to questions such as how can growth be translated into community development and well-being.

Furthermore, recently highlighted questions about what we are actually aiming to sustain in sustainable tourism (other than tourism!) and what makes communities resilient to unsustainable socio-economic and environmental changes are the key challenges for sustainable tourism development in the future (Saarinen 2014; McCool & Bosak 2016). These questions and issues are not simple, as tourism–community relations form a complex system characterized by constant change, occasional turbulence, and the interplay between social and ecological elements on various scales. Related challenges are multiscalar by nature, covering research, policy-making and governance needs, and new business models. In order to understand the key elements in benefit creation, such as how tourism generates impacts and sustainable benefits for communities, there is a need to create theoretical frameworks that would provide context-sensitive models for community benefit creation in sustainable tourism planning and development. This is a highly demanding task. However, even in the predicted era of uncertainty (see Kohl & McCool 2016), there is a need to focus on knowledge-based policies that aim to guide and influence the future sustainability of host–guest relations in tourism.

References

Arnstein, S. R. (1969), 'A ladder of citizen participation', *Journal of the American Planning Association*, **35**(4): 216–24.

Blaikie, P. (2006), 'Is small really beautiful? Community-based natural resource management in Malawi and Botswana', *World Development*, **34**: 1942–57.

Bourdieu, P. (1986), 'The forms of capital', in J. Richardson (ed.), *Handbook of Theory and Research for the Sociology of Education*, pp. 241–58, New York: Greenwood.

Bramwell, B. (2011), 'Governance, the state and sustainable tourism: a political economy approach', *Journal of Sustainable Tourism*, **19**(4–5): 459–77.

Bramwell, B. and B. Lane (2011), 'Critical research on the governance of tourism and sustainability', *Journal of Sustainable Tourism*, **19**(4–5): 411–21.

Burns, P. (1999), 'Paradoxes in planning: tourism elitism or brutalism?', *Annals of Tourism Research*, **26**(2): 329–48.

Butler, G. and C. Rogerson (2016), 'Inclusive local tourism development in South Africa: evidence from Dullstroom', *Local Economy*, **31**(1–2): 264–81.

Butler, R. (1980), 'The concept of a tourist area life cycle of evolution: implications for management of resources', *The Canadian Geographer*, **24**(1): 5–12.

Butler, R. (1999), 'Sustainable tourism: a state-of-the-art review', *Tourism Geographies*, **1**(1): 7–25.

Butler, R. (2006), *The Tourism Area Life Cycle: Applications and Modifications*, Tonawanda, NY: Channel View Publications.

Butler, R. and T. Hinch (eds.) (2007), *Tourism and Indigenous Peoples*, Oxford: Butterworth-Heinemann.

Church, A. and T. Coles (eds.) (2007), *Tourism, Power and Space*, Abingdon: Routledge.

Daly, H. (1996), *Beyond Growth: The Economics of Sustainable Development*, Boston, MA: Beacon Press.

Doxey, G. V. (1975), 'A causation theory of visitor-resident irritants: methodology and research inferences', in *The Impact of Tourism, The Travel Research Association, Sixth Annual Conference Proceedings, San Diego, California, September 8–11*, pp. 195–8.

Espiner, S., C. Orchiston and J. Higham (2017), 'Resilience and sustainability: a complementary relationship? Towards a practical conceptual model for the sustainability–resilience nexus in tourism', *Journal of Sustainable Tourism*, **25**(10): 1385–1400.

Goffman, E. (1959), *The Presentation of Self in Everyday Life*, New York: Anchor Books.

Gursoy, D., C. G. Chi and P. Dyer (2009), 'Locals' attitudes toward mass and alternative tourism: the case of Sunshine Coast, Australia', *Journal of Travel Research*, **649**(3): 381–94.

Hall, C. M. (1994), *Tourism and Politics: Policy, Power and Place*, Chichester: Wiley & Sons.

Hall, C. M. (2008), *Tourism Planning*, Harlow: Prentice-Hall.

Hall, C. M. (2011), 'Policy learning and policy failure in sustainable tourism governance: from first- and second-order to third-order change?', *Journal of Sustainable Tourism*, **19**(4–5): 649–71.

Hall C. M., S. Gössling and D. Scott (eds.) (2015), *The Routledge Handbook of Tourism and Sustainability*, Abingdon: Routledge.

Harvey, D. (2005), *A Brief History of Neoliberalism*, Oxford: Oxford University Press.

Holden, A. (2018), 'Environmental ethics for tourism—the state of the art', *Tourism Review*, https://doi.org/10.1108/TR-03-2017-0066.

Holling, C. S. (1973), 'Resilience and stability of ecological systems', *Annual Review of Ecology and Systematics*, **4**: 1–23.

Hunt, C., W. Durham, L. Driscoll and M. Honey (2015), 'Can ecotourism deliver real economic, social, and environmental benefits? A study of the Osa Peninsula, Costa Rica', *Journal of Sustainable Tourism*, **23**(3): 339–57.

Hunt, C. and A. Stronza (2014), 'Stage-based tourism models and resident attitudes towards tourism in an emerging destination in the developing world', *Journal of Sustainable Tourism*, **22**(2): 279–98.

Hunter, C. (1995), 'On the need to re-conceptualize sustainable tourism development', *Journal of Sustainable Tourism*, **3**(3): 155–65.

IUCN (1980), *World Conservation Strategy: Living Resource Conservation for Sustainable Development*, Gland: IUCN.

Jamal, T. (2013), 'Resiliency and uncertainty in tourism', in A. Holden and D. Fennell (eds.), *The Routledge Handbook of Tourism and Environment*, pp. 505–20, London: Routledge.

Jamal, T. and A. Stronza (2009), 'Collaboration theory and tourism practice in protected areas: stakeholders, structuring and sustainability', *Journal of Sustainable Tourism*, **17**(2): 169–89.

Jessop, B. (2004), 'Hollowing out the "nation-state" and multilevel governance', in P. Kenneth (ed.), *A Handbook of Comparative Social Policy*, pp. 11–25, Cheltenham: Edward Elgar Publishing.

Jessop, B. (2010), 'Government and governance', in A. Pike, A. Rodriquez-Pose and J. Tomaney (eds.), *Handbook of Local and Regional Development*, pp. 239–48, London: Routledge.

Jones, S. (2005), 'Community-based ecotourism: the significance of social capital', *Annals of Tourism Research*, **32**(2): 303–24.

Kavita E. and J. Saarinen (2016), 'Tourism and rural community development in Namibia: policy issues review', *Fennia*, **194**(1): 79–88.

Kim, K., M. Uysal and M. J. Sirgy (2013), 'How does tourism in a community impact the quality of life of community residents?', *Tourism Management*, **36**: 527–40.

Kohl, J. and S. F. McCool (2016), *The Future Has Other Plans: Planning Holistically to Conserve Natural and Cultural Heritage*, Golden, CO: Fulcrum Publishing.

Latour, B. (2015), 'Telling friends from foes in the time of the Anthropocene', in C. Hamilton, F. Gemenne and C. Bonneuil (eds.), *The Anthropocene and the Global Environmental Crisis: Rethinking Modernity in a New Epoch*, pp. 145–55, Abingdon: Routledge.

Lew, A. A. (2014), 'Scale, change and resilience in community tourism planning', *Tourism Geographies*, **16**(1): 14–22.

Li, W. J. (2006), 'Community decision-making: participation in development', *Annals of Tourism Research*, **33**(1): 132–43.

Liu, Z. (2003), 'Sustainable tourism development: a critique', *Journal of Sustainable Tourism*, **11**: 459–75.

Ma, M. and R. Hassink (2013), 'An evolutionary perspective on tourism area development', *Annals of Tourism Research*, **41**(1): 89–109.

MacCannell, D. (1976), *The Tourist: A New Theory of the Leisure Class*, New York: Schoken Books.

McCool, S. F. (2009), 'Constructing partnerships for protected area tourism planning in an era of change and messiness', *Journal of Sustainable Tourism*, **17**(2): 133–48.

McCool, S. F. and K. Bosak (2016), *Reframing Sustainable Tourism*, Berlin: Springer.

McCool, S. F., N. Moisey and N. Nickerson (2001), 'What should tourism sustain? The disconnect with industry perceptions of useful indicators', *Journal of Travel Research*, **40**: 124–31.

McKercher, B., B. Prideaux, C. Cheunga and R. Law (2010), 'Achieving voluntary reductions in the carbon footprint of tourism and climate change', *Journal of Sustainable Tourism*, **18**(3): 297–317.

Martin, R. and P. Sunley (2014), 'On the notion of regional economic resilience: conceptualization and explanation', *Journal of Economic Geography*, **15**(1): 1–42.

Mathieson, A. and G. Wall (1982), *Tourism: Economic, Physical and Social Impacts*, London: Longman.

Mbaiwa, J. E. (2005), 'The sociocultural impacts of tourism development in the Okavango Delta, Botswana', *Journal of Tourism and Cultural Change*, **2**(3): 163–85.

Moore, A. (2015), 'Tourism in the Anthropocene Park? New analytic possibilities', *International Journal of Tourism Anthropology*, **4**(2): 186–200.

Mosedale, J. (2014), 'Political economy of tourism', in A. A. Lew, C. M. Hall and A. M. Williams (eds.), *The Wiley Blackwell Companion to Tourism*, pp. 55–65, Chichester: John Wiley & Sons.

Murphy, P. (1985), *Tourism: A Community Approach*, London: Methuen.

Nofre, J., G. Giordano, A. Eldridge, J. Martins and J. Sequera (2018), 'Tourism, nightlife and planning: challenges and opportunities for community liveability in La Barceloneta', *Tourism Geographies*, **20**(3): 377–96.

Novelli, M. and K. Gebhardt (2007), 'Community-based tourism in Namibia: "reality show" or "window dressing"?', *Current Issues in Tourism*, **10**(5): 443–79.

Nunkoo, R. and D. Gursoy (2012), 'Residents' support for tourism: an identity perspective', *Annals of Tourism Research*, **39**(1): 243–68.

Nunkoo, R. and H. Ramkissoo (2011), 'Developing a community support model for tourism', *Annals of Tourism Research*, **38**(3): 964–88.

Nunkoo, R. and H. Ramkissoo (2012), 'Power, trust, social exchange and community support', *Annals of Tourism Research*, **39**(2): 997–1023.

Nunkoo, R. and S. L. J. Smith (2013), 'Political economy of tourism: trust in government actors, political support, and their determinants', *Tourism Management*, **36**: 120–32.

Ostry, J. D., P. Loungani and D. Furceri (2016), 'Neoliberalism oversold?', *Finance & Development*, **53**(2): 38–41.

Othman, S., A. Ramli and A. S. Ahmad (2017), 'The conceptual framework on social entrepreneurship activities as a mediator between social capital and the performance of Malaysian small and medium enterprises (SMEs)', *Journal of Education and Social Sciences*, **8**(1): 110–14.

Paasi, A., E. K. Prokkola, J. Saarinen and K. Zimmerbauer (eds.) (2019), *Borderless Worlds for Whom? Ethics, Moralities, Mobilities*, Abingdon: Routledge.

Pearce, P. L., G. Moscardo and G. F. Ross (1996), *Tourism Community Relationships*, Oxford: Pergamon/ Elsevier Science.

Pike, A., A. Rodríguez-Pose and J. Tomaney (2007), 'What kind of local and regional development and for whom?', *Regional Studies*, **41**(9): 1253–69.

Poon, A. (1993), *Tourism, Technology and Competitive Strategies*, Wallingford: CABI International.

Putnam, R. D. (1995), 'Bowling alone: America's declining social capital', *Journal of Democracy*, **6**(1): 65–78.

Putnam, R. D. (2001), *Bowling Alone: The Collapse and Revival of American Community*, New York: Simon and Schuster.

Reed, M. (1997), 'Power relations and community-based tourism planning', *Annals of Tourism Research*, **24**: 566–91.

Reid, M. (2003), *Tourism, Globalization and Development: Responsible Tourism Planning*, London: Pluto Press.

Rhodes, R. A. W. (1996), 'The new governance: governing without government', *Political Studies*, **44**: 652–67.

Rogerson, C. M. and J. Saarinen (2018), 'Tourism for poverty alleviation: issues and debates in the global south', in C. Cooper, S. Volo, W. C. Gartner and N. Scott (eds.), *The SAGE Handbook of Tourism Management: Applications of Theories and Concepts to Tourism*, pp. 22–37, London: Sage.

Saarinen, J. (2010), 'Local tourism awareness: community views in Katutura and King Nehale conservancy, Namibia', *Development Southern Africa*, **27**: 713–24.

Saarinen, J. (2014), 'Critical sustainability: setting the limits to growth and responsibility in tourism', *Sustainability*, **6**(11): 1–17.

Saarinen, J. and M. Lenao (2014), 'Integrating tourism to rural development and planning in the developing countries', *Development Southern Africa*, **31**(3): 363–72.

Saarinen, J., C. Rogerson and C. M. Hall (2017), 'Geographies of tourism development and planning', *Tourism Geographies*, **19**(3): 307–17.

Scheyvens, R. (1999), 'Ecotourism and the empowerment of local communities', *Tourism Management*, **20**: 245–9.

Scheyvens, R. (2002), *Tourism for Development: Empowering Communities*, Harlow: Prentice Hall.

Scheyvens, R. (2011), *Tourism and Poverty*, London: Routledge.

Schwab, K. (2015), 'The fourth industrial revolution', *Foreign Affairs*, retrieved October 4, 2018 from https://www.foreignaffairs.com/articles/2015-12-12/fourth-industrial-revolution.

Sharpley, R. (2000), 'Tourism and sustainable development: exploring the theoretical divide', *Journal of Sustainable Tourism*, **8**: 1–19.

Sinclair, T. (1998), 'Tourism and economic development: a survey', *Journal of Development Studies*, **34**(5): 1–51.

Smith, V. (ed.) (1989), *Hosts and Guests: The Anthropology of Tourism* (2nd edn), Philadelphia: University of Pennsylvania Press.

Stoker, G. (1998), 'Governance as theory: five propositions', *International Social Science Journal*, **50**(155): 17–28.

Telfer, D. J. and R. Sharpley (2008), *Tourism and Development in the Developing World*, Abingdon: Routledge.

Timothy, D. J. (1999), 'Participatory planning: a view of tourism in Indonesia', *Annals of Tourism Research*, **26**(2): 371–91.

Tosun, C. (1999), 'An analysis of the economic contribution of inbound international tourism in Turkey', *Tourism Economics*, **5**(3): 217–50.

Tosun, C. (2000), 'Limits to community participation in the tourism development process in developing countries', *Tourism Management*, **21**(6): 613–33.

Tosun, C. (2006), 'Expected nature of community participation in tourism development', *Tourism Management*, **27**(3): 493–504.

UNCTAD (2017), *A Commitment to Inclusive Trade*, Geneva: UNCTAD.

UNEP (United Nations Environment Programme) (2005), *Making Tourism More Sustainable: A Guide for Policy Makers*, Paris: UNEP.

United Nations Secretary-General's High-Level Panel on Global Sustainability (2012), *Resilient People, Resilient Planet: A Future Worth Choosing, Overview*, New York: United Nations.

UNRISD (United Nations Research Institute for Social Development) (2015), *Revisiting Sustainable Development*, Geneva: UNRISD.

Urry, J. (1990), *The Tourist Gaze: Leisure and Travel in Contemporary Societies*, London: Sage.

Utting, P. (2015), 'Foreword', in P. Utting (ed.), *Revisiting Sustainable Development*, Geneva: UNRISD.

Valentin, A. and J. H. Spangenberg (2000), 'A guide to community sustainability indicators', *Environmental Impact Assessment Review*, **20**: 381–92.

van der Duim, R., D. Meyer, J. Saarinen and K. Zellmer (eds.) (2012), *New Alliances for Tourism: Conservation and Development in Eastern and Southern Africa*, Delft: Eburon.

WCED (United Nations World Commission on Environment and Development) (1987), *Our Common Future*, Oxford: Oxford University Press.

World Bank (2012), *Transformation through Tourism: Development Dynamics Past, Present and Future*, Washington, DC: World Bank.

World Bank Group (2017), *World Bank Annual Report 2017*, Washington, DC: World Bank.

World Tourism Organization (1993), *Sustainable Tourism Development: Guide for Local Planners*, Madrid: UNWTO.

World Tourism Organization (2016), *UNWTO Annual Report 2015*, Madrid: UNWTO.

14 Managing environmental impacts of tourism

Teresa Cristina Magro-Lindenkamp and Yu-Fai Leung

Introduction

Tourism, like other industries or human activities, can induce varying degrees of negative environmental and social consequences. The convergence of fast-growing global tourism, largely due to higher income, more leisure time and lower transportation costs, allied with technological advancements and new tastes in recreation activity, has raised serious concerns about tourism's sustainable future. "Overtourism" is now a trendy new term that frames such concerns about tourism's environmental and social impacts (CNN Travel 2018a; Manjoo 2018).

Research on tourism's environmental impacts, the subject matter of this chapter, has confirmed that tourism is responsible for a host of direct and indirect environmental effects (Blumstein et al. 2017; Rankin, Ballantyne & Pickering 2015). Less certain, however, is whether this body of scientific literature actually informs or shapes tourism development toward greater sustainability. Indeed, as is evident in the twenty-first century, while many scientific reports are published on the environmental impacts of tourism, there is concurrently significant and growing marketing and promotion of new destinations for tourism consumption each year, and tourists keep rushing to their dream destinations to check off their "bucket list."

The infinite flood of visitors cannot be accommodated at the world's most popular destinations. The result is alteration of historic cities, loss of natural and cultural characteristics and other negative consequences presented in this chapter. We cannot blame only technology and the rise of low-cost airlines, as cited in Manjoo (2018) and CNN Travel (2018a)—social media and blogging sites, like Instagram, YouTube, Twitter and Facebook are as much the culprit. Besides popular destinations like Venice, Rome, Barcelona, Amsterdam, Bali and Rio de Janeiro, tourists now are choosing places based on how photogenic and attractive these places are when posted on social media. Photos have always been taken in touristic places, but they can now be posted at light speed, thanks to the combination of the Internet and smartphones.

These new developments change the way how tourist hotspots are seen by people. Most of the impressions posted on social media are a "good news show." Few tourists will relate negative experiences to their friends back home as they don't like to admit or confess the failure of their dream holiday destination. Their behavior changes as their main concern is to present only happiness and satisfaction, even if those feelings are not truly felt—most are involved in taking pictures rather than experiencing the site itself. In summer 2018, this behavior resulted in a fight between tourists as they were trying to get the best spot for a selfie in front of Trevi Fountain (Piazza di Trevi), in Rome, as reported by international media.[1]

Experience is the tourism product, and comfort contributes significantly to the quality of the contemporary tourist experience. It is a market that tries to sell to a select public who can afford it, an environment that provides a distinct experience of everything that has already been seen or experienced using terms such as "different," "beautiful" and "amazing," often in a superlative way. When this place no longer seems to exercise the previously held seduction, like primitivism, structures are added that promise new and spectacular experiences. Add to that sensory aspects and physical comfort—in extremely hot places all indoor environments are provided with air conditioning and very cold areas with comfortable heaters. Cafeterias are replaced by gourmet restaurants, camping becomes glamping, and the comfort of the urban environment is brought to former rural areas, or even primitive ones. This would not be a problem if the process did not trigger the search for new areas and new infrastructure that changed the landscape and brought significant habitat losses with a high cost of restoring natural conditions.

Tourism can be a driving force for positive impacts, like the establishment of protected areas, environmental restoration, species conservation and economic contribution to local communities. Tourism is often pointed out by all nations as a source of employment. Its importance was recognized beyond the economic factor as the United Nations (UN) designated 2017 as the International Year of Sustainable Tourism for Development (IUCN 2017), underscoring the contributions of the tourism sector to global development based on the three pillars of sustainability: economic, social and environmental. While these benefits are widely celebrated, it is worthwhile to ask: Whom does tourism serve? Who are the beneficiaries? If we compare tourism with other industries, do we have a specific group that only works with tourism or are investors related to other businesses that end up causing environmental damage? Who consumes this product that is often presented as a consumer's dream?

According to McCool (this volume, Chapter 1) the twentieth century was characterized by the important aspects of sustainability and sustainable development. The approaches were built on the basis of a tripod that action needed to be taken to address economic viability, acceptability and environmental responsibility. Also, according to Bosak and McCool (this volume, Chapter 15), rather than thinking about how we can sustain tourism perhaps we should be asking "*what it is that*

1 See https://edition.cnn.com/travel/article/selfie-fight-trevi-fountain/index.html.

tourism should sustain." Should tourism sustain benefits and profits? Or should the emphasis be on the quality of tourist products, experiences and consumptions? Or its role as an effective conservation and community development tool?

This chapter illustrates the current issues and challenges of environmental impacts of tourism with international examples, and explores research gaps and questions for the twenty-first century as informed by a system-thinking approach.

Tourism's environmental impacts: a wicked problem?

According to the UN World Tourism Organization, international tourist arrivals worldwide amounted to 1.32 billion in 2017, and are expected to reach 1.8 billion by 2030 (UNWTO 2018). We can see that many of them wander the world in search of unique experiences but end up joining thousands in the same destinations, compromising the quality of their own experience and reducing the environmental quality of those destinations. Places like Machu Picchu, Venice, the Great Wall of China, the Taj Mahal and the Grand Canyon, among so many others, have attracted large agglomerations of visitors and there is a great chance that they are not having a positive experience in these places.

Protected areas, due to their scenic beauty, natural resources, cultural heritage and outdoor recreation opportunities, are popular destinations for tourists who are interested in connecting with nature. Protected area visitation has been reported inconsistently across countries so it is difficult to know the precise number. Recent work by Balmford et al. (2015) estimated that there were about eight billion visits to protected areas globally, with 80 percent occurring in Europe and North America. In the United States, the national park system received a record high of 330 million visits in 2017 (NPS 2018) while the state park system received over 807 million visits in the 8,292 state park units across 50 states (NASPD 2018). In the coming years, protected area visitation will likely grow due to increasing interest in nature and culture, along with rising numbers of outbound tourists from populous countries with a strong economy, such as China and India. Because of the magnitude of impacts that tourism has, and the conservation values of protected area systems and networks, a variety of guidelines have been developed with the hope of managing tourism and visitor activities to reduce negative environmental impacts while maximizing positive influences (Leung et al. 2018).

In making a relative analysis of the expansion in interest in outdoor recreation activities and tourism growth, we could be surprised that this increase may not be so significant everywhere. Even though visits to national parks and other protected areas have increased (in this case we evaluated the United States and Brazil), we can infer that the highest concentration occurs in those places with expressive infrastructure. The number of visitors seeking a genuine primitive experience has not increased significantly. This is evident in an analysis of visitor's to Brazilian National Parks, which increased by 20 percent in 2017, to 10.7 million. The major concentration is in Tijuca

National Park (RJ) with 3.3 million in 2017 (compared with 2.7 million in 2016). The Iguaçu National Park is the second most visited, with 1.8 million tourists in 2017 (1.6 million in 2016), followed by the Jericoacoara National Park and the Fernando de Noronha Marine Park. All these areas offer good infrastructure for visitors.

Research has also shown that the relationship between the number of tourists and the negative environmental impacts is not always linear (Hammitt, Cole & Monz 2015). Thus, considering the non-linearity between nexus, what worries researchers are growing habitat losses and fragmentation associated with newly "discovered" tourist areas, increasing use of motorized equipment in primitive areas, and litter and sewage discharged from large cruise ships that act as a mobile urban concentration, leading pathogens and non-degradable materials to the oceans.

The negative environmental impact directly caused by increasing numbers of visitors at the same place could be softened by an appropriate treatment of the sewage and limited consumption. Nevertheless, it is precisely the marketing of products that has been stimulated by tourism, making it more profitable as an investment. Consumption is directly related to wear, loss, extraction and exploitation and obeys the market forces of supply and demand. So, on top of the presence of tourists and the environmental costs of transportation by air, land and sea, we have the environmental costs of overexploitation on products sold as souvenirs. We are not discussing the positive effects on the economy and value for the locals of selling their products. What matters is that earnings and profits by tourism mostly don't remain in the area but flow to places where the investment comes from. There is an increased extraction of materials for making souvenirs, which causes damage—for example, semiprecious stones, seashells containing living creatures (live shell examples ensure more beauty), wood for handicrafts, and other products of plant and animal origin. It is quite difficult to change this logic of high consumption linked to tourism as it is part of the economic model of the market. Therefore, new research needs to reconsider the relation and balance between jobs and the protection of natural resources.

Besides tourism in urban and rural areas, wilderness may suffer further damage. In the future, areas that remain primitive will also suffer pressure for use of natural resources besides the use for visiting purposes. We already deal with existing pressure in natural areas in, for example, mining, wood extraction and heavy weaponry used on military ranges, among others. These activities have far more impacts on nature than tourism, but this does not mean that we shouldn't discuss the collateral damages of tourism. A deep analysis of its negative effects is necessary and urgent to ensure a natural heritage for future generations.

In addition to these examples is a third variable (influence factor) that acts on the proposed model, the financial one. Some activities represent a surplus expenditure that the majority of the population cannot afford. For this group there is a huge volume of package deals by coach or airplane. To make destinations more attractive, the tourist agencies portray the places as paradises (natural and urban), with the perfect climate, with the perfect light, without insects. Thus, the public, mostly fast viewers of a scenario, look at the horizon, take a picture, and hurry to continue with the sequence that was sold to them in the package deal. They are landscape consumers.

Three significant limiting factors that are present in the search for tourist destinations can be avoided or mitigated: access, which was previously cited as a technique for managing the number of visitors, is more efficient and faster; the fear factor has become a safe adrenaline rush; and the economic issue, which has made greater access possible for a population to the benefits of tourism with some social achievements.

We assume that all forms of tourism have the potential to cause environmental impacts. The criteria for this judgment are based on the fact that there is alteration of the environment and fauna displacement if the environment is primitive or semi-primitive. Another finding relates to the fact that part of the impacts originate in the tourist's behavior. Once on vacation the attitude (the behavior) of human beings usually changes due to the euphoria we feel: we consume more, eat more. Suddenly we are averse to rules. After all, we are on vacation to relax. We want to experience the different, and that has to be so special, so fantastic, to the point of becoming an event to be narrated as a good story for our friends and relatives. And above all, the landscape in the background has to be unique and photogenic. It seems that with each new decade tourists are more eager for more exciting experiences. Not only do we travel more distantly from home, but also more times a year. We go further and further, deeper and deeper, higher and higher. Into the oceans, caves, deserts and mountains, and with more comfort each year.

We are convinced that visitors and locals create a better understanding of undesirable environmental impacts once they undergo these inconveniences. Talking about negative environmental and social effects, what should be decisive in the decision-making process for a decision-maker? Apparently, social impacts in tourist areas are less accepted. The negative environmental effects usually occur in the medium and long term and usually are not perceived by visiting tourists. People asked about their experiences often relate noise and garbage as negative impacts on their experience, but they do not perceive that factors like excessive noise have negative influences on reproductive success of the avifauna, or that litter contaminates water sources. Even for tour operator these impacts are not visible. In the case of very popular urban centers, like Venice, Barcelona, New York, Paris, Amsterdam, Bali and Rio de Janeiro, local citizens show their annoyance with the so-called overtourism, and in some cases hostility breaks out. In big cities the demand for overnight stays are causing problems—tourists using Airbnb make unwelcome noise in residential areas and the growth of this form of overnight stay decreases the housing supply for locals and provokes higher prices.

In the case of natural, primitive or semi-primitive environments, there is a need for managers to find ways to assess the negative effects, especially in low-resilient areas. According to McCool (this volume, Chapter 1), "Wicked problems occur in situations where scientists disagree on cause–effect relationships and social agreement on goals does not exist. And because of these situations we need to look carefully at the suitability and usefulness of the paradigms currently used to guide research." These situations perfectly describe the challenges of tourism's environmental impacts because they are influenced by a host of social and environmental factors with unclear causal relationships. Social consensus among stakeholders on the desired level and types of tourism seldom exists either. The "floating cities" phenomenon is a case in point.

The floating cities: a case of a wicked problem

One of the big problems is the floating cities, giant cruise ships moving in fragile areas like the lagoons of Venice, Italy. Savio (2018) disagreed strongly that tourism is a major benefit to Venice and its people. Contrary to the often-stated advantages to the local economy brought by visiting cruise ships, they are not as great as alleged. Savio mentions the big impacts on the lagoon microsystem in Venice. He also says that most jobs are temporary, seasonal and low-paid. According to Savio, Venice has lost much of its identity in order to adapt to the demands of tourists.

In order to maintain its financial success, large-scale tourism is increasing newfound destinations and unfortunately, most of them have low environmental resilience. Tourism activities in urban areas also cause disturbing environmental effects. Mainly it is a matter of water pollution and emission of pollutants. Probably the most shocking examples are the large cruise liners that pour tons of organic and non-organic material into the oceans every year—it is estimated that one billion gallons (3.8 billion liters) are pumped into the ocean annually. This sewage teems with bacteria, heavy metals, pathogens, viruses, pharmaceuticals, and more substances that are potentially harmful to animals and humans (Avellaneda et al. 2011). But the waste produced in cruise liners can be transformed into energy that can be used on the ship itself. The technology exists to produce biogas by transforming every form of kitchen waste into a source of energy, first into gas, which in turn can be used for electricity generation to supply heating and power. The Dutch company Enki Energy works with projects that replace non-renewable fossil fuels with organic waste streams (Blankenborg 2018). According to the UK director of the trade body Cruise Lines International Association, the sector has an interest in maintaining the cleanliness of the oceans and is investing in technologies for water treatment and recycling as resort hotels use on land (McVeigh 2017), but implementation takes a long time.

Another relevant impact refers to air quality. "German environment group Nabu claims one medium cruise ship emits as many pollutants as five million cars going the same distance. It says the ships belch out 3,500 times more sulphur dioxide than cars—although international rules to reduce sulphur emissions in shipping are due to come into force in 2020" (McVeigh 2017).

The floating cities that receive this great flux of tourists do not always have sewage treatment systems and garbage collection sized for these large flows of people in a seasonal way. Resizing these systems, in turn, can represent expenses that municipalities may not be willing to pay.

There are cases of cities that have increased their capacity to receive tourists with the construction of hotels and restaurants as close as possible to the shoreline. These buildings disfigure and, in some cases, destroy the landscape all in order to please the tourist. Very often the hospitality industry uses the landscape to its utmost to promote their products and attract clients.

Emerging trends, new impacts?

On top of persistent environmental issues associated with mass tourism as described above, emerging trends in recreation and tourism are generating concerns because they can exacerbate current impacts while causing new impacts to the environment. Here are just two examples.

One emerging trend is that what was unique and unusual used to have a strong attraction for a small number of people due to the real risk of the activity. This ensured that climbing high mountains, entering a labyrinth of caves, or diving deep in the ocean, going as far as possible in inhospitable surroundings, were activities pursued by few people. Even the real and imminent possibility of death resulting from these types of activity was a positive and attractive factor. Many of these activities, which were dangerous and life-threatening decades ago, are nowadays relatively safe and comfortable although they still take place in inhospitable environments. Thus, the value that these environments represent for personal development has changed. Everything can be done with great security but one has to pay a high price for these activities. The extreme example of this is orbital space tourism. This activity may be for a select happy few, but it comes with the penalties of high emissions of carbon dioxide and a relatively significant effect on climate change.

Another emerging trend is trail running, adventure racing, and other competition-based sporting events and associated tourism that are rapidly gaining popularity in both protected areas and cities in different continents (Newsome 2014). The rising demand is partly due to changing public interests and effective marketing of non-profit or for-profit event organizers while protected area agencies are increasingly embracing public and physical health objectives to demonstrate their societal relevance. However, intensive use of trails, facilities and natural areas by large numbers of people in these events can cause significant negative impacts on natural resources and recreation infrastructure. Some examples include vegetation damage, soil erosion, informal trail formation, wildlife disturbance, and trash (Ng et al. 2018). Protected area managers are challenged to weigh the social and economic benefits of these events with the environmental and social costs. One critical question is when and where these events (as well as their size and type) are appropriate.

Unfortunately, we still do not have a lot of knowledge on new forms of recreation and tourist activities (e.g., organized sport events). But new projects will find some helpful answers in current recreation ecology knowledge that could assist in building political will and courage.

Currently, one of the issues to be addressed is to make more evident to decision-makers, both public and private, that the natural environment is an important resource, essential for maintaining ecosystem services. This also has a great value for public health and understanding how ecosystems function. These elements together constitute a strong argument that can be used to justify the existence and creation of protected areas around the globe.

What do we already know about environmental impacts of tourism?

Environmental impacts of tourism have been a subject of investigation since the 1960s and a body of literature has been developed (Buckley 2011, 2018; Holden & Fennell 2013; Newsome, Moore & Dowling 2012). One of the earlier compilations, "The ecological impacts of outdoor recreation on mountain areas in Europe and North America" (Bayfield & Barrow 1983), brought together results of research involving experimentation and bibliographical revision with the objective of assisting governmental decision-making related to the tourist use of natural areas. At that time, the rapid increase of nature-based recreation and tourism, mostly in the forms of walking, climbing and nature study, and associated environmental impacts, had been recognized in North America and Europe. Emerging activities and the use of new technologies at the time was also discussed, including hang-gliding, microlight planes, ballooning, cross-country vehicles, damage from motor bikes, sub-aqua sports, power boating, cross-country skiing, helicopter skiing, and so on (Bayfield & Barrow 1983, p. i).

During the ensuing four decades, numerous studies have been conducted by researchers in different disciplines who have attempted to understand the extent, nature, patterns or processes of tourism impacts on all ecological components. As a result, a wide variety of tourism impacts have been identified and quantified, including soil and vegetation damage, wildlife disturbance and water quality degradation (Buckley 2011; Hall and Lew 2009; Newsome, Moore & Dowling 2012). These impacts can be found in both mass tourism and ecotourism destinations (Blumstein et al. 2017), although there are greater concerns about the threats of these tourism-induced stresses to more fragile landscapes and ecosystems geared towards nature-based tourism, ecotourism and geotourism. The accumulation of literature has afforded a systematic review of past research at larger geographic scales (e.g., Pickering & Mount 2010; Rankin, Ballantyne & Pickering 2015), as well as the development of generalized frameworks or models for predicting impact intensity and disturbance patterns (Gutzwiller, D'Antonio & Monz 2017; Martin & Butler 2017; Monz, Pickering & Hadwen 2013). Recent research on tourism impacts have expanded to all world regions (Barros, Monz & Pickering 2015; Cunha & Costa 2018; Leung 2012; Magro-Lindenkamp & Passold 2018; Tejedo et al. 2016; Yang et al. 2014). The latest studies also utilize technologies, such as GPS tracking, Geographic Information Systems (GIS) and social media data, in improving our understanding on spatial and temporal patterns of tourism impacts at the landscape level (Monz et al. 2016; Walden-Schreiner et al. 2018; Yang et al. 2014).

While the body of literature on this topic is well established for North America, Australia and Europe, fewer published literature exists to inform management in other continents. Research is always lagging behind emerging activity trends and new technologies, such as organized sporting events and the use of drones, to name just two.

System thinking for solving a wicked problem

As summarized above, a good amount of research has been done on environmental impacts of tourism. Although the results of the surveys indicate negative effects on the environment, judging their relevance and choosing the actions to deal with this consequence will have to be done by people—an analysis and judgment that involves personal and collective values beyond scientific knowledge—an example of a "wicked problem."

Thus, from what was developed in the twentieth century and in the face of the uncertainties of the twenty-first century (this volume, McCool, Chapter 1), it is important to re-evaluate judgments, taking into account new knowledge and new values. This may seem difficult but if the decision-makers opt for the use of a balance that favors long-term maintenance of environmental quality, following the laws and parameters of environmental quality established for each region, it is a good beginning. In the case of environmental impacts, the answer will come from accepting yes or no options to the alteration of ecological systems. But it is very difficult to leave human judgment aside—it will always present an interference factor. At the same time, determining if ecological changes are 100 percent due to recreational tourist activities or natural phenomena is a very complex issue. Consequently, it is also difficult to make concrete regulations for tourist activities. One example is bird watching and the use of playback. Some studies on bird watching indicate that the activity is responsible for increased nest predation and nest abandonment, changes in reproductive behavior, induced habituation and changes in vocal behavior (e.g., Harris and Haskell 2013; Mennill, Ratcliffe & Boag 2002), but positive arguments are also presented (Sekercioglu 2002).

The tourist industry is presenting itself as an ecological clean sector which benevolently improves access to interesting places and creates more opportunities for people to appreciate the environment, learn, and to grow personally. But are their motives so genuine? Do the trips around the globe contribute to a higher understanding among people and nations? Are consumption patterns during travels decreasing or at least staying the same, once we begin to understand our relationship with the environment? We also have to pay attention to the question if improving access for the tourist industry brings benefits to the local population in terms of enhancing quality of life, access to formal education, public health and increased income.

So called "progress" comes with improved access, as often stated, and better access goes hand in hand with other major changes in local behaviors. But better access doesn't necessarily mean that everybody judges this as positive progress. A tourist destination can suffer from a lower evaluation and appreciation due to a large influx of tourists, cultural disfiguration and loss of the natural environment (more hotels, more asphalt, more paved stairs and so on). What has attracted people before the implanted progress ceases to exist in terms of quality, at least for a certain category of tourists. This may lead to the prospect of a search for new potential destinations with unique characteristics, and might risk devaluation of old destinations and a loss of return on investments.

In order to grab the interest of the tourist and to hold them on the spot for as long as possible (resulting in money being spent in places that are no longer considered as authentic) investment is needed, and usually a substantial amount—new attractions and services need to be implanted. In this case, nature and local habitats are not given priority. We are mainly talking about areas that still maintain the characteristics of primitive environments, with high conservation values and large numbers of environmental services.

Considering the actual current models, we should assume that negative interference in ecosystem services alone should represent an impediment to the implementation of a tourism activity or the temporary suspension of activity already in progress.

This might sound like a very radical proposal that would have serious consequences for all parties involved, especially where the economy is based exclusively on tourist activity. In many cities, the quality of life of its residents is under pressure due to the high level of visitors. The most extreme example is perhaps Venice—with a population of just 50,000 inhabitants the city hosts 30 million visitors annually, and groans under efforts to sustain a 24-hour economy. There is very little time to rest as tourists continue to arrive in what is effectively a year-long season. In other examples, places like Amsterdam and Barcelona are discussing how to contain the growing inflow of tourists all year round, and Maya Bay on the island of Koh Phi Phi Leh (where Leonardo DiCaprio famously starred in the movie *The Beach*) was closed in 2018 in an attempt by the authorities to reverse the damage to marine life over previous decades. According to CNN Travel (2018b), at the end of September 2018, Thailand's Department of National Parks, Wildlife and Plant Conservation announced that the bay will remain closed indefinitely until the ecosystem has recovered completely. This kind of drastic action has already been taken at other tourist sites also in Thailand, such as the Koh Tachai beach that was closed in 2016 for environmental recovery.[2] In April 2018, President Duterte of the Philippines closed down the island of Boracay which receives 2 million visitors per year. He blamed the tourist industry for dumping sewage in the open sea and said that the sector had changed the island into an open cesspool (Maas 2018). Finally, in Peru, new strategies have been embraced by the public sector to deal with the large numbers of tourists visiting the old Inca citadel of Machu Picchu. The Ministry of Culture issued a raft of rules in July 2017 that aimed to protect the Inca site by modifying visitation practices (Peru Guide 2017; Sachs 2018)—for example, one can now choose to visit either in the morning (6 a.m. to noon) or the afternoon (noon to 5:30 p.m.).

Finding solutions: management innovations and research gaps

Tourism is indeed an important industry for the generation of wealth, and it contributes to development and understanding among different peoples. The term "sustainable tourism," however, is still very much related to trips made in natural areas or ecotourism. When we talk about sustainable tourism we immediately think that this will happen in natural and primitive areas. But sustainable tourism should be a practice in all types of environments, including urban areas, and with greater respect for the local inhabitants. The "Good Agricultural Practices" guide is a prime example of what can be achieved—it even includes advice on hygienic practices in packing plants.[3]

The commitment of promoting sustainability in tourism must be a shared goal between decision-makers in both the public and private domains, representatives of the tourism trade and travelers themselves. Tourists usually do not search for a trip because of sustainability considerations. He or she is seduced most of the time by the destination and the price of the package. The first initiative may come from the tourism trade itself, if it desires to remain sustainable in the future. In general, holiday agencies do not report on the negative effects of travel. One significant contribution that could be made would be reporting the amount of CO_2 emissions involved in each activity and service. Regarding transport, accommodation and food, it would be ideal if the tourist could select services and activities that have less impact on the environment. For example, instead of motorized vehicles one could walk or rent a bicycle; local produce could be consumed, rather than pre-packaged, commercially produced food imported to the region; and train transport is to be preferred over airplanes because of their lower harmful emissions.

A good example is the Chilean Atacama Desert area. The extremely dry region is inhospitable for agriculture or raising cattle, but the dramatic landscape and the Indian culture are perfect for developing tourist activities. In a span of twenty

2 See https://www.theguardian.com/world/2016/may/17/thailand-closes-koh-tachai-andaman-sea-island-to-tourists-coral-reefs.

3 See http://www.fao.org/3/a-a1193e.pdf.

years, the sleepy town of San Pedro de Atacama has become an ecotourism center of international fame. Arguably, the village has undergone significant changes that have decharacterized the original community, but the residents (who were generally poor until the 1980s) are now safe from poverty thanks to ecotourism and investment from other regions in Chile and abroad. Further studies on the region would be welcome to establish if there have been negative impacts on nature and the indigenous population. However, sustainable tourism is still not an option for a large part of the global population and tourist hotspots often enter a cycle of use and destruction. One example is the beaches in northeast Brazil such as Morro de São Paulo (Bahia), and Jericoacoara and Canoa Quebrada, in Ceará state. Originally, all of the sites were villages inhabited by poor fishermen, but once discovered by hippies, backpack tourism started to flourish. Years later the first hostels were built by locals, followed by hotels constructed by external investors to attract richer tourists. With international fame rising, backpack tourism ended and backpackers started to look for unspoiled beaches, thus starting the cycle anew.

At some point this cycle needs to change. The question is where to draw the line between (eco)tourism which keeps the ambience as pure as possible, and high-density tourism to ensure comfortable earnings for the agencies and hotels, and a satisfactory return for investors. The starting point for discussion and policy-making is that all kind of tourism must, first and foremost, be sustainable.

Management innovations and practices that address environmental impacts of tourism for the twenty-first century can benefit from further research, even though a substantive body of literature exists. There are persistent and recurring research needs. For example, it is always crucial to understand how to better manage tourism as a productive activity by identifying the most fragile sites from the environmental point of view. Science should be able to tell where tourism can be most beneficial and where it is most negative. This is more urgent in the case of natural and primitive areas with low resistance.

As mentioned before, it is extremely difficult to change the logic of high consumption and tourism once it is part of the economic model of the market. A recurring research gap is how to find the balance between economic benefits such as job creation with the protection of natural resources. New research needs to reconsider the relation and balance between jobs and the protection of natural resources.

Beyond the persistent research needs, the system-thinking approach (this volume, McCool, Chapter 1) offers new insights on research questions, which we should ask if environmental impacts of tourism are to be better understood and effectively managed in the twenty-first century:

- How do environmental stresses induced by tourism at the site level, such as facility development and wildlife harassment, lead to ecosystem- and landscape-scale consequences? What is the spatial and temporal lag of these consequences?

- How would tourism management actions, such as use limits or activity restrictions, in one destination change visitation and tourist behavior in adjacent destinations?
- What are the effects of changing transport modes on tourist distribution patterns, their interactions with nature and the local community, and the nature of their experiences?
- How do climate change effects, such as changing phenology of flowing plants and disturbance such as wildfires, affect spatial and temporal patterns of visitation demand, tourism operation and visitor experiences? What are the effects of these changes on local communities?

Concluding remarks

This chapter has summarized a number of contemporary tourism trends that are shaping the intensity and types of environmental impacts. We have also discussed examples of management innovations and practices addressing tourism's environmental impacts. Our discussion and formulation of research questions have benefitted from a system-thinking approach, which considers tourism's site- and destination-level environmental problems in the context of a larger scale system involving adjacent communities and destinations as well as different stakeholders.

Science already reveals the negative effects of tourism on the environment, but there has not yet been effective action taken in management decision-making (often disregarded), based on the knowledge generated in the last 40 years. Analyzing the results of research related to environmental impacts of tourism, we have verified that the linearity between causal links is not a constant rule. There are other parameters that influence the negative impact on the natural environment, such as the method of traveling, environmental resilience and resistance. If this was true for the twentieth century, one can ask what is missing from decision-making processes that needs to be directed to ensure efficient and low-impact tourism management in the twenty-first century. In the vast majority of cases, there is a lack of courage regarding decision-making based on the scientific knowledge generated so far.

According to Magro-Lindenkamp and Passold (2018, p. 14) tourism is a great opportunity for society to recognize natural ecosystems and their associated values and also a chance for local economic development. The Millennium Ecosystem Assessment (2005) acknowledges social-cultural benefits generated by humans interacting with their natural environment. Recreation is currently considered as one of the significant ecosystem services in natural areas. Finally, in natural areas there are ecological and social conditions understood by society and once threatened they need to be protected. Because the natural attributes and values could be lost at some level by touristic use, the same attributes and values should always guide new research and monitoring programs.

References

Avellaneda, Pedro M., Englehardt, James D., Olascoaga, Josefina, Babcock, Elizabeth A., Brand, Larry, Lirman, Diego, Rogge, Wolfgang F., Solo-Gabriele, Helena and Tchobanoglous, George (2011), 'Relative risk assessment of cruise ships biosolids disposal alternatives'. *Marine Pollution Bulletin*, **62**: 2157–69.

Balmford, Andrew, Green, Jonathan M. H., Anderson, Michael, Beresford, James, Huang, Charles, Naidoo, R. Robin, Walpole, Matt and Manica, Andrea (2015), 'Walk on the wide side: estimating the global magnitude of visits to protected areas'. *PLoS Biology*, **13**(2): e1002074.

Barros, Augustina, Monz, Christopher and Pickering, Catherine (2015), 'Is tourism damaging ecosystems in the Andes? Current knowledge and an agenda for future research'. *Ambio*, **44**: 82–98.

Bayfield, Neil G. and Barrow, Graham C. (eds) (1983), 'The ecological impacts of outdoor recreation on mountain areas in Europe and North America'. *Recreation Ecology Research Group. R.E.R.G. Report No. 9*.

Blankenborg, Stefan (2018), 'One World publiceert over de broodvergister van Enki Energy in Amersfoort', retrieved May 14, 2018 from http://www.enki-energy.com/en/.

Blumstein, Daniel T., Geffroy, Benjamin, Samia, Diogo S. and Bessa, Eduardo (2017), *Ecotourism's Promise and Peril: A Biological Evaluation*. New York: Springer.

Buckley, Ralf (2011), 'Tourism and environment'. *Annual Review of Environment and Resources*, **36**(3): 1–20.

Buckley, Ralf (2018), 'Managing the natural environment for tourism', in C. Cooper, S. Volo, W. C. Gardner, and N. Scott (eds), *The SAGE Handbook of Tourism Management*, pp. 1–9. London: Sage.

CNN Travel (2018a), 'Can the world be saved from overtourism?', retrieved October 11, 2018 from http://www.edition.cnn.com/travel/article/overtourism-solutions/index.html.

CNN Travel (2018b), 'Thailand bay made popular by "The Beach" closes indefinitely', retreived October 13, 2018 from http://www.edition.cnn.com/travel/article/maya-bay-closure-thailand/index.html.

Cunha, Andre A. and Costa, Cassio M. M. (2018), 'Nature tourism research in Brazil: a preliminary scientometric approach of the last 20 years', in Andre A. Cunha, Teresa C. Magro and Stephen F. McCool (eds), *Tourism and Protected Areas in Brazil: Challenges and Perspectives*, pp. 25–45. New York: Nova Science Publishers.

Gutzwiller, Kevin, D'Antonio, Ashley L. and Monz, Christopher A. (2017), 'Wildland recreation disturbance: broad-scale spatial analysis and management'. *Frontiers in Ecology and the Environment*, **15**(9): 517–24.

Hall, Colin M. and Lew, Alan A. (2009), *Understanding and Managing Tourism Impacts: An Integrated Approach*. New York: Routledge.

Hammitt, William E., Cole, David N. and Monz, Christopher A. (2015), *Wildland Recreation: Ecology and Management*. 3rd edn. Chichester: Wiley Blackwell.

Harris, J. Berton C. and Haskell, David G. (2013), 'Simulated birdwatchers' playback affects the behavior of two tropical birds'. *PLoS ONE*, **8**(10): 1–8.

Holden, Andrew and Fennell, David A. (2013), *Routledge Handbook of Tourism and the Environment*. London: Routledge.

IUCN (International Union for Conservation of Nature) (2017), '2017 International year of sustainable tourism for development', retrieved April 19, 2018 from http://www.tourism4development2017.org/about.

Leung, Yu-Fai (2012), 'Recreation ecology research in East Asia's protected areas: redefining impacts?' *Journal for Nature Conservation*, **20**(6): 349–56.

Leung, Yu-Fai, Spenceley, Anna, Hvenegaard, Glen T. and Buckley, Ralf (eds) (2018), 'Tourism and visitor management in protected areas: guidelines for sustainability' (Best Practice Protected Area Guidelines Series No. 27). Gland: IUCN.

McVeigh, Tracy (2017), 'As British tourists take to the seas, giant cruise ships spread pollution misery', retrieved May 19, 2018 from https://www.theguardian.com/environment/2017/jan/08/ports-pollution-cruising-ships-freight-sea.

Maas, Michel (2018), 'Boracay gehoorzaamt de president: het paradijselijke strand is leeg', retrieved October 13, 2018 from https://www.volkskrant.nl/nieuws-achtergrond/boracay-gehoorzaamt-de-president-het-paradijselijke-strand-is-leeg~befb4d7e/.

Magro-Lindenkamp, Teresa C. and Passold, Anna J. (2018), 'Coping with the effects of tourism in natural areas', in Andre A. Cunha, Teresa C. Magro and Stephen F. McCool (eds), *Protected Areas and Tourism in Brazil: Challenges and Perspectives*, pp. 1–24. New York: Nova Science Publishers.

Manjoo, Farhad (2018), '"Overtourism" worries Europe: how much did technology help get us there?', retrieved October 11, 2018 from https://www.nytimes.com/2018/08/29/technology/technology-overtourism-europe.html.

Martin, R. and Butler, D. R. (2017), 'A framework for understanding off-trail trampling impacts in mountain environments'. *George Wright Forum*, **34**(3): 354–67.

Mennill, Daniel J., Ratcliffe, Laurene M. and Boag, Peter T. (2002), 'Female eavesdropping on male song contests in songbirds'. *Science*, **296**: 873.

Millennium Ecosystem Assessment (2005), *Ecosystems and Human Well-being: Synthesis.* Washington, DC: Island Press. Retrieved July 18, 2019 from https://www.millenniumassessment.org/documents/document.356.aspx.pdf.

Monz, Christopher, D'Antonio, Ashley, Lawson, Steve, Barber, Jesse and Newman, Peter (2016), 'The ecological implications of visitor transportation in parks and protected areas: examples from research in US national parks'. *Journal of Transport Geography*, **51**: 27–35.

Monz, Christopher, Pickering, Catherine and Hadwen, Wade L. (2013), 'Recent advances in recreation ecology and the implications of different relationships between recreation use and ecological impacts'. *Frontiers in Ecology and the Environment*, **11**(8): 441–6.

NASPD (National Association of State Park Directors) (2018), *Annual Information Exchange—Statistical Report of State Park Operations: 2016–2017*. Raleigh, NC: NASPD.

Newsome, David (2014), 'Appropriate policy development and research needs in response to adventure racing in protected areas'. *Biological Conservation*, **171**: 259–69.

Newsome, David, Moore, Susan A. and Dowling, Ross K. (2012), *Natural Area Tourism: Ecology, Impacts and Management* (2nd edn). Bristol: Channel View Publications.

Ng, Sai-Leung, Leung, Yu-Fai, Cheung, Suet-Yi and Fang, Wei (2018), 'Land degradation effects initiated by trail running events in an urban protected area of Hong Kong'. *Land Degradation and Development*, **29**(3): 422–32.

NPS (U.S. National Park Service) (2018), 'National Park system sees more than 330 million visits', retrieved October 15, 2018 at https://www.nps.gov/orgs/1207/02-28-2018-visitation-certified.htm.

Peru Guide (2017), 'Machu Picchu: new entrance rules, July 1, 2017', retrieved October 13, 2018 from www.theonlyperuguide.com/machu-picchu-new-entrance-rules-july-01-2017/.

Pickering, Catherine and Mount, Ann (2010), 'Do tourists disperse weed seed? A global review of unintentional human-mediated terrestrial seed dispersal on clothing, vehicles and horses'. *Journal of Sustainable Tourism*, **18**(2): 239–56.

Rankin, Ben, Ballantyne, Mark and Pickering, Catherine (2015), 'Tourism and recreation listed as a threat for a wide diversity of vascular plants: a continental-scale review'. *Journal of Environmental Management*, **154**: 293–8.

Sachs, Andrea (2018), 'Peru devises new rules to deal with ever growing Machu Picchu crowds', retrieved February 7, 2018 from https://www.chicagotribune.com/lifestyles/travel/ct-machu-picchu-tourism-rules-20180205-story.html.

Savio, Roberto (2018), 'Tourism should be regulated, before it is too late', retrieved May 8, 2018 from www.ipsnews.net/2018/01/tourism-regulated-late/.

Sekercioglu, Cagan H. (2002), 'Impacts of birdwatching on human and avian communities'. *Environmental Conservation*, **29**(3): 282–9.

Tejedo, Pablo, Benayas, Javier, Cajiao, Daniela, Albertos, Belén, Lara, Francisco, Pertierra, Luis R., Andres-Abellan, Manuela, Wic, Consuelo, Lucianez, Maria J., Enríquez, Natalia, Justel, Ana and Reck, Gunther K. (2016), 'Assessing environmental conditions of Antarctic footpaths to support management decisions'. *Journal of Environmental Management*, **177**: 320–30.

UNWTO (UN World Tourism Organization) (2018), 'Tourism highlights. 2018 edition', retrieved October 27, 2018 from https://www.e-unwto.org/doi/book/10.18111/9789284419876.

Walden-Schreiner, Chelsey, Rossi, Sebastian D., Barros, Agustina, Pickering, Catherine and Leung, Yu-Fai (2018), 'Using crowd-sourced photos to assess seasonal patterns of visitor use in mountain-protected areas'. *Ambio*, **47**(7): 781–93.

Yang, Mingyu, van Coillie, Frieke, Liu, Min, de Wulf, Robert, Hens, Luc and Ou, Xiaokun (2014), 'A GIS approach to estimating tourists' off-road use in a mountainous protected area of Northwest Yunnan, China'. *Mountain Research and Development*, **34**(2): 107–17.

15 A research agenda for sustainable tourism: some ideas worth pursuing

Keith Bosak and Stephen F. McCool

In this book, we have noted that sustainable tourism presents not only many benefits to the global citizenry but many challenges as well. The challenges are wicked and messy, which in a practical sense means that the conventional problem solving of the past no longer works. And what we have seen throughout the text is that the diversity of challenges that confront sustainable tourism is enormous; we and our authors have illustrated only some of these challenges. Our purpose was to provide a peek into much of the literature of sustainable tourism.

In spite of the many, many definitions of sustainable tourism we are really not sure what it is we are referring to. As such, Chapter 1 briefly suggested that we focus on outcomes rather than what type of tourism it may be; thus, because of its messiness, we will only know sustainable tourism when we see it, a conundrum produced by the twenty-first century, one of turbulence and uncertainty. We do feel, however, that sustainable tourism is generally not amenable to the reductionist and simplistic scientific practice of the past. We need to think and act more holistically than we have before.

This is not to say that research of the past 20–30 years has yielded little knowledge. Indeed, that research has helped us understand bits and pieces of knowledge about components of what we may call, as Bosak stated, a "tourism system," which is many interconnected and nested systems, depending upon the question being pursued. Numerous studies have yielded information and knowledge that is useful and actionable to decision-makers. And the authors in this volume have provided some wonderful summaries of that research.

However, we need to think differently about tourism that is sustainable into the future as implied in the first three chapters. The past has also yielded ways of thinking that are no longer useful in resolving the wicked and messy challenges of the future. For example, despite an immense amount of research on resident attitudes toward tourism in communities for at least the last 25–30 years, European tourism marketers were apparently caught off-guard by the "overtourism" rebellion of 2016. And in North America, managers were surprised by the outcry against high levels of visitor use in national parks, many of which had sponsored so-called "carrying capacity" studies in the not-so-distant past. In both cases, some kind of failure in the research–manager interface occurred.

We need to do better through policy-relevant research, effective communication and useful capacity building. The book summarizes some ways and approaches that will help us in the coming years, an important goal as international travel is scheduled to rise by about 50 percent between now (2019) and 2030.

Returning to systems thinking

The predicted increase in international travel can be viewed as beneficial to local and national economies as it will contribute to labor income as well as foreign exchange, lifting personal incomes and making the economies of some states more viable. But it can also be viewed as a crisis in the making, because of potential for the negative unanticipated consequences it brings (as suggested by Magro-Lindenkamp and Leung, Chapter 14)—everything from congested transportation to clogged cities to jammed parks, each of which carries surprises we cannot predict at the moment (that is why they are called surprises).

These are surprises in large part because while we have learned much about how each component of a tourism system works we have neither learned as much as we need to nor do we hold any substantial understanding of the *relationships among tourism components* of complex, adaptive systems. This is a major challenge in sustainable tourism because different actors are involved in the varying components. For example, an airline may provide the fast, inexpensive and convenient transportation to a destination, while a government may build the airport, a local airport authority allocates landing spots, and a different agency may manage the airport. And then a local tour operator may provide bus services, a restaurant is engaged to handle meals, another company provides lodging, while some other entity manages an attraction, such as a national park, under a totally different set of rules and policies. Do we really understand how all these components relate to each other? Probably not. Do we know where the leverage points are located? Maybe. Do we understand how actions at higher levels of systems affect each component and how they ripple through this system? Not really.

It is a difficult system to think about, conceptualize or apply, as noted by Bosak (Chapter 2) and Espiner, Higham and Orchiston (Chapter 3). Tourism has many components, which makes it complicated. It is wicked and messy. It is non-linear, so a little bit of change in one component may result in a large amount of change in another. That makes it complex. Its emergent property, we believe, is sustainability and resilience. So, we have to ask, are these components in a tourism system explicitly and integratively contributing to those two emergent properties, just as a steering wheel, engine and transmission work together as three of the component properties to produce an automobile whose emergent property is extreme mobility?

This is an important agenda item worthy of the most competent of science—practitioner teams. The outcome of this kind of research is not so much knowledge per se, but rather understanding, because it tells us how a system works, where its feedback loops are located, where the leverage points and delays are located, how tightly causes and effects are coupled, and how resilient it may be when subject to perturbations.

Toward understanding visitor experiences

To a large degree, the visitor experience is our product, the goal of the system we need to better understand how it works. We have learned much, as shown by Miller et al. (Chapter 5), in the last three decades about the motivations people hold for different attractions and places, their expectations for a visit and their evaluations of experiences, as well as how the contributions of site attributes, social interactions and norms, infrastructure and context affect these domains. This is why people spend money to visit Venice, vacation in Yellowstone National Park or travel to the deepest corners of the Amazon basin to have adventure, interact with indigenous people or hike into a tropical forest.

If the parts of the system are not synchronized it is unlikely a good experience will happen, outside of those situations where luck is involved. Sometimes the system is "home built" as when we are the purchaser, organize our own air reservations, our surface transportation, food, lodging and other services we desire, but many people have neither the skill nor the time nor the information to do so. And so, we rely on many others to synchronistically create an opportunity for highly valued experience. We have all had these experiences.

But sometimes the synchronism doesn't happen, and an experiential disaster occurs. We may even totally miss our chance or sometimes it is merely a short delay or something similar. So, how robust or resilient is tourism to such disturbances? Can the visitor recover from a missed flight or a December snowstorm? To the traveler, these are not insignificant events. Can the appropriate industry sector help the traveler recover?

As noted by Gianna Moscardo (Chapter 6), marketing is not solely focused on promotion, but encompasses making connections between the traveler and destination, so this system is essential to good, even transformative experiences, experiences that become a lifetime memory. We do need to know more about these connections, how to manage them and how to make them more effective.

On engaging communities looking at tourism as a development strategy

We have often noted that communities, whether a large metropolis or a village of a few families located in a remote Himayalan region, is where tourism really occurs, whether they are the destination, or "basecamp" for travelers seeking attractions located in rural or wildland areas. This note directly relates to the question of *what it is that tourism should sustain.* It could be small businesses, a villager livelihood, or resilience of the community. It is a question for public and community engagement, but too often villagers may not even know about this question, implicitly giving their faith to a scientist who knows little of the particular social or cultural context.

Knowing the response to this question is essential as part of a community-based vision. Scientists such as Saarinen (Chapter 13) have several roles here; for example, providing information about consequences of tourism development that might occur, or developing a framework for a community discussion about this, or suggesting what kind of information may be useful (such as who to market to, how marketing connections will be made, and by whom). Who bears the costs, who receives the benefits, and how benefits are distributed are significant challenges that need to be debated and discussed before decisions are made about the rationale for community-based tourism.

Business plays a key role here as Kelly Bricker noted in Chapter 9. Business and business-like approaches are needed to reduce the risk of failure not because an idea is not financially feasible but because caring for a business, particularly a start-up, requires attention to detail, an understanding of the value chain, and long, long hours of work. Science can help identify attractive tourism ideas, build understanding of the market, and suggest means of efficiently monitoring consequences before a problem becomes insurmountable.

Spenceley and Rylance (Chapter 8) pointed out how science can help communities and states pursue the UN Sustainable Development Goals (SDGs) if they so wish to do so. These goals are related principally to the economic, social, environmental and health domains. One or more of these goals may be the response to *what it is that tourism should sustain.* For example, the UN SDG 3 focuses on good health, which Derrien, Cerveny and Wolf discussed in Chapter 10. As they pointed out, there are considerable positive health effects of recreation and tourism. The simple activity of walking, either on city streets looking at historic architecture or through beautiful wilderness landscapes, can improve cardiovascular fitness or reduces stress. So science plays several roles here.

Tourism changes communities sometimes toward the negative, sometimes toward the positive. The community of Kimmswick in Missouri mentioned by McCool (Chapter 1) is an example of a community where tourism saved the town's infrastructure at the least and gave viable economic options to its citizens. In other places, such as Venice, this has happened as well, but it grew too large over a long period of time for the residents to accommodate.

Science plays an important role here, not just about the economic consequences of tourism development, but also about what it does to its culture and power relationships within the community. We have often seen an underlying assumption that residents of smaller communities are homogeneous, but they are not. Residents are not equally impacted by tourism development—some gain economically, some don't, and this often results in jealousy or conflict. Science can help understand how money flows through a community, and it can also help understand cultural and social consequences.

Managing nature-based attractions

About 15 percent of the terrestrial surface of our world is protected as formally designated areas, much of it in places, such as national parks, where little human development is permitted and in other areas where natural resources providing food, sustenance and shelter are carefully managed. These places are significant locations for nature-based activities: who has not wanted to hike a forest trail, play on a deserted beach in Brazil, gaze at the wonders of the Namib Sand Sea, see an Asian elephant, or simply desire a picnic in one of the beech forests of Europe?

As our cities, tourism and recreation visits have grown remarkably in the last two decades and in places that are sensitive to human-induced impacts, ranging from soil erosion to devegetation and loss of wildlife. And yet people still seek to come, because, like the monuments of Europe and Asia, they want to know something about their natural heritage before it is lost, although many of them would hope that, like the editors and authors of this volume, their grandchildren will live to experience it.

Cerveny and Miller (Chapter 11) summarized the challenges facing sustainable tourism in these places through an international survey. These are considerable barriers facing managers of protected areas as budgets for both management and research have declined, almost concurrently with the rise of interest in sustainable tourism. It is quite an incredible challenge, made more arduous in that we have little understanding of sustainable tourism for nature-based activities, few managers to sustain some of the most special places on earth, and less knowledge and even less understanding of how to provide opportunities for people to learn about the natural systems upon which life is contingent.

And like many of the other challenges observed in this volume, it is wicked and messy. Protecting these places, most of which occur on public land, is connected to education, which is related to income and wealth, which is linked to governance, and which is coupled with . . . you name it. We cannot solve one problem without solving another. All these variables are loosely coupled, which means there are delays between each one, like between achieving a secondary education and holding political office or some other civic responsibility.

Human and ecological systems come together most notably in nature-based tourism. The uncertainty of this social-ecological system requires management to monitor conditions and consequences as pointed out most graphically by Thomsen et al. in Chapter 12. Monitoring provides valuable feedback, and can help avoid those surprises we mentioned briefly earlier in this chapter. Avoiding surprises can also be viewed as a strategy to reduce costs in the long run.

Understanding impacts

Another theme that is not new to sustainable tourism research but still of utmost importance is understanding impacts. All tourism produces impacts and those impacts ripple through the system in both space and time. Therefore, research on impacts of tourism is still important but requires new approaches that take a deeper look at them and how they act as disturbances (good or bad) in the socio-ecological systems of which tourism is a part. Understanding impacts is the first step in building adaptive capacity and ultimately resilience. However, just trying to understand impacts can be messy. As Scott (Chapter 7) and Loehr, Addinsall and Weiler (Chapter 4) point out with reference to climate change, its impacts can be far-ranging and can positively and/or negatively impact a destination. Therefore, understanding the context within which impacts occur is also important as are the power dynamics that produce certain impacts, particularly negative impacts. Moscardo (Chapter 6) alludes to this in her discussion of overtourism as a backlash against the numerous negative impacts of tourism in many locations. She notes that these unanticipated impacts have arisen from unsustainable practices.

Often, negative attitudes towards tourism such as those expressed in the overtourism movement are an outcome of a lack of benefits from tourism for those that live in destinations. Overtourism also represents negative impacts to the tourist experience through crowding and congestion and, as Miller et al. (Chapter 5) point out, requires understanding the visitor experience from both the impacts of management of a destination on visitors, and the impact visitors have on that destination that affect the visitor experience, an example of a system with obvious feedback.

Impacts are not entirely negative and much of the focus of sustainable tourism research has been on understanding and enhancing the positive impacts. Spenceley and Rylance (Chapter 8) point out the importance of understanding the positive impacts of tourism in contributing to Sustainable Development Goals through poverty alleviation, social justice and enhancing well-being. Bricker (Chapter 9) continues the discussion with a deep dive into business models that could enhance positive economic, ecological and social impacts. Derrien, Cerveny and Wolf (Chapter 10) point out the impacts on human health from tourism both on tourists and in destinations. Finally, Saarinen (Chapter 13) highlights tourism impacts and how they might translate into benefits for communities.

Concluding thoughts

Our thoughts about the challenges covered in this book are almost as diverse as tourism itself and reflect a broad array of issues that need attention from researchers. However, it is important to note that while these topics are treated as separate for the sake of organization in this book, they are in fact overlapping and interconnected. Communities, businesses and governments all interact in destinations that are being impacted by global economies, climate change and changing visitor preferences. At the same time, these destinations are striving to produce quality visitor experiences while staying true to their values and conserving their natural heritage. Governments and NGOs promote sustainable tourism as a development strategy that can alleviate poverty, support community health and wellness, and contribute to biodiversity conservation. We want to point out that there are myriad connections, intersections and relations in our world, in tourism and in the research we conduct to understand and improve the sustainability of tourism.

This is why we have placed such an emphasis on systems thinking, embracing complexity and the need for research that spans disciplines and moves beyond reductionist thinking. Tourism as a complex adaptive system is self-organizing and characterized by emergent properties and non-linear feedbacks. These characteristics reflect the world we live in today that is changing at an ever increasing rate and becoming more interconnected and complex. Tourism like any other system, is subject to shocks and disturbances whether from economic downturns, political instability, natural disasters or even the changing preferences of visitors. Actors in the system need the capacity to adapt to disturbances in a way that maintains or strengthens resilience but can only do so with good information.

These challenges are somewhat similar to the editors' preferences and experiences for outdoor recreation. We choose a destination and activity for which we want to have an experience: Keith's desire to backcountry ski down a local mountain or Steve's much less adventurous trip to a local cross-country ski area. We each build information as we go, knowing more each trip about the best spots for powder for Keith, and the best place for a powerful snowy landscape photo for Steve. On each trip we adapt, because we know more and better prepare for the conditions, which sometimes are different to what is anticipated, and we are thus surprised. We adapt to the surprise, take a slightly different route, monitoring snow conditions, weather, temperature, visitor numbers, and change our route accordingly. Most times we come back with a really great experience because we monitor, compare to our desires and capabilities, adapt and are happy. Sometimes we don't have a good experience. But the results are taken account of in the next round of a skiing experience, no matter how adventurous.

And so this is what we do. Understanding the current state of tourism and all of its components can assist researchers, practitioners and communities in anticipating, preparing for and adapting to disturbances and shocks, thus building resilience. It is our hope that this volume will help to set a direction for future research by addressing some of the salient topics in sustainable research while also providing a conceptual foundation to address these topics in a way that leads to a deeper understanding and, ultimately, sustainability in all types of tourism.

Index